ISSN 0044-2267

ZEITSCHRIFT FÜR ANGEWANDTE MATHEMATIK UND MECHANIK

INGENIEURWISSENSCHAFTLICHE FORSCHUNGSARBEITEN

BAND 61 1981 HEFT 6

AUS DEM INHALT

A K A D E M I E - V E R L A G · B E R L I N

ZAMM Bd. 61 Nr. 6 S. 209—276 Berlin, Juni 1981

EVP 18,— M 34115

INHALT

Hauptaufsätze

Kleine Mitteilungen

Wir bitten, **Manuskriptsendungen** zweifach (Original und eine Kopie, sprachlich einwandfrei, Formeln mit Maschine oder in Druckschrift geschrieben) an folgende Anschrift zu richten:

Zeitschrift für Angewandte Mathematik und Mechanik,
Institut für Mechanik der Akademie der Wissenschaften der DDR
DDR-1199 Berlin, Rudower Chaussee 5.

Zu den Arbeiten, die als Hauptaufsätze bestimmt sind, ist auf gesondertem Blatt eine Zusammenfassung von 5 bis 10 Zeilen in englischer und (möglichst) deutscher und russischer Sprache beizufügen. Ausführliche Hinweise für die Autoren, um deren strikte Berücksichtigung gebeten wird, finden sich im Anschluß an das Inhaltsverzeichnis des Jahrganges **60** (1980).

Die Autoren erhalten von den Hauptaufsätzen 75, von den Kleinen Mitteilungen 25 Sonderdrucke ohne Berechnung, darüber hinaus weitere Sonderdrucke gegen Berechnung.

Der Verlag behält sich für alle Beiträge das Recht der Vervielfältigung und Übersetzung vor.

Bestellungen sind zu richten

- **in der DDR** an den Postzeitungsvertrieb unter Angabe der Kundennummer des Bestellenden oder an den AKADEMIE-VERLAG, DDR-1080 Berlin, Leipziger Straße 3—4
- **im sozialistischen Ausland** an eine Buchhandlung für fremdsprachige Literatur oder an den zuständigen Postzeitungsvertrieb
- **in der BRD und Berlin(West)** an eine Buchhandlung oder an die Auslieferungsstelle KUNST UND WISSEN, Erich Bieber OHG, D-7000 Stuttgart 1, Wilhelmstraße 4—6
- **in den übrigen westeuropäischen Ländern** an eine Buchhandlung oder an die Auslieferungsstelle KUNST UND WISSEN, Erich Bieber GmbH, Ch-8008 Zürich/Schweiz, Dufourstraße 51
- **im übrigen Ausland** an den Internationalen Buch- und Zeitschriftenhandel; den Buchexport, Volkseigener Außenhandelsbetrieb der Deutschen Demokratischen Republik, DDR-7010 Leipzig, Postfach 160; oder an den AKADEMIE-VERLAG, DDR-1080 Berlin, Leipziger Straße 3—4

ZEITSCHRIFT FÜR ANGEWANDTE MATHEMATIK UND MECHANIK

Herausgeber und Chefredakteur: Prof. Dr. Günter Schmidt. Redaktion: Dr. Winfried Heinrich, Dr. Horst Weinert, Dipl.-Math. Friedhild Dudel, Helga Rühl, Institut für Mechanik der Akademie der Wissenschaften der DDR. Verlag: Akademie-Verlag, DDR-1080 Berlin, Leipziger Straße 3—4; Fernruf: 2236221 oder 2236229. Telex-Nr.: 114420; Bank: Staatsbank der DDR, Berlin, Kto.-Nr.: 6836-26-20712. Anschrift der Redaktion: Institut für Mechanik der Akademie der Wissenschaften, DDR-1199 Berlin, Rudower Chaussee 5, Fernruf: 6702841/3639 und 3643.

Veröffentlicht unter der Lizenznummer 1282 des Presseamtes beim Vorsitzenden des Ministerrates der Deutschen Demokratischen Republik. Gesamtherstellung: VEB Druckerei „Thomas Müntzer", 5820 Bad Langensalza. Erscheinungsweise: Die Zeitschrift für Angewandte Mathematik und Mechanik erscheint monatlich. Die 12 Hefte eines Jahres einschließlich Tagungsheft bilden einen Band. Bezugspreis je Band 360,— M zuzüglich Versandspesen (Preis für die DDR 216,— M). Bezugspreis je Heft 30,— M (Preis für die DDR 18,— M). Bestellnummer dieses Heftes: 1009/61/6.

AN (EDV) 35937

ISSN 0044-2267

ZAMM 61, 209 – 214 (1981)

S. I. Pai / S. Menon / Z. Q. Fan

Strong Shock Wave Propagation in a Mixture of a Gas and Dusty Particles with Gravitational Force

Beim Zusammenstoß eines Staubkometen mit einem Planeten entstehen starke Stoßwellen im staubbeladenen Gas der Koma. In diesem Fall spielt die Gravitationskraft eine wesentliche Rolle. Bei Berücksichtigung der Gravitationskraft existieren keine Ähnlichkeitslösungen für die Ausbreitung starker Stoßwellen. In dieser Arbeit untersuchen wir den Einfluß der Gravitationskraft in einem staubbeladenen Gas für den Fall großer Froude-Zahlen mit Hilfe der Störungstheorie. Es werden numerische Lösungen für Geschwindigkeits-, Dichte- und Druckverteilungen hinter dem Stoß und für die Stoßgeschwindigkeiten sowie ihre räumlich-zeitliche Anordnung unter verschiedenen Bedingungen angegeben. Die Gravitationskraft hat einen dominierenden Einfluß auf die Strömungsfelder bei der Ausbreitung starker Stoßwellen in einem staubbeladenen Gas.

When a dusty comet collides with a planet, strong shock waves are produced in the dusty gas of the coma. In this case, the gravitational force plays an important role. With gravitational force, similarity solution [1] of the strong shock wave propagation does not exist. In this paper, we study the effect of the gravitational force in a dusty gas by perturbation theory for the case of large Froude number. Numerical solutions are presented on the velocity, density and pressure distributions behind the shock as well as the shock speeds and its location with time under various conditions. The gravitational force does have predominant influence on the flow fields of the strong shock propagation in a dusty gas.

При столкновении кометы, которая состоит из пыли, с планетой возникают сильные ударные волны в газе комы, которы содержит пыль. В этом случае гравитационная сила играет важную роль. При решении задачи с учетом гравитационной силы подобные решения для распространения сильных ударных вольн не существуют. В этой работе исследуется влияние гравитационной силы в газе, который содержит пыль, для случая больших чисел Фруда с помощью методов теорий возмущения. Данны численные решения для распределения скорости, плотности и давления за ударом, для скоростей удара как и пространственно-временное распределение этых величин при различных условиях. Гравитационная сила имеет определяющее влияние на полие течения при распространении сильных ударных вольн в газе, который содержит пыль.

1. Introduction

In a previous paper, [1] we considered the similarity solutions of a strong shock wave propagating in a mixture of a gas and dusty particles without gravitational forces. One of the applications of the strongshockwave propagation analysis is the cometary impact on planetary surface [2]. In such an astrophysical problem, the gravitational force may not be neglected. In this paper we consider the effect of gravitational force on the strong shock wave propagation in a mixture of a gas and dusty particles. With the presence of gravitational force, we do not have similarity solutions even when we consider the case of equilibrium flow with constant initial density distribution. However, when the shock wave is strong with large value of shock speed the gravitational force effect will be small because the Froude number $Fr = D^2/y_2g$ will be very large. Here, $y_2(t)$ is the location of the shock front and may be used as the characteristic length of our problem. Thus we may use the perturbation theory based on the similarity solution of Reference [1] to find the first-order effect of gravitational force on the strong shock wave propagation in a mixture of a gas and dusty particles.

2. Fundamental Equations and Important Non-Dimensional Parameters

Similar to Reference [1], we consider the one-dimensional unsteady equilibrium flow of a mixture of a gas and small solid particles but with the gravitational force terms included.

The unknowns of our problem are

(i) velocity of the mixture, $u(y, t)$
(ii) temperature of the mixture, $T(y, t)$
(iii) pressure of the mixture, $p(y, t)$,
(iv) density of the mixture, $\varrho(y, t)$,

where y is the spatial coordinate and t is the time.

The fundamental equations for our problem are:

(i) *Equation of continuity*

$$\frac{\partial \varrho}{\partial t} + \frac{\partial \varrho u}{\partial y} = 0 . \tag{1}$$

(ii) *Equation of motion*

$$\varrho \left(\frac{\partial u}{\partial t} + u \frac{\partial u}{\partial y} \right) = - \frac{\partial p}{\partial y} + \delta_1 \varrho g \tag{2}$$

where g is the magnitude of the gravitational acceleration, $\delta_1 = 1$ if the direction of the coordinate y is the same as that of the gravitational force, and $\delta_1 = -1$ if the direction is opposite to that of the gravitational force. The velocity $u(y, t)$ is in the same direction as the spatial coordinate y.

(iii) *Equation of energy*

$$\frac{\partial}{\partial t}\left[\varrho\left(U_m + \frac{1}{2}u^2 + \Phi_m\right)\right] + \frac{\partial}{\partial y}\left[\varrho u\left(U_m + \frac{1}{2}u^2 + \Phi_m\right) + \frac{p}{\varrho}\right] = 0 \tag{3}$$

where U_m is the internal energy of the mixture per unit mass and $\Phi_m = -\delta_1 yg$ is the potential energy measured from reference altitude, $y = 0$.

(iv) *Equation of state*

$$p = \frac{1 - k_p}{1 - Z}\varrho RT \tag{4}$$

where R is the gas constant, Z is the volume fraction of the solid particles in the mixture and k_p is the mass concentration of the solid particles in the mixture.

The internal energy of the mixture depends on the mass concentration k_p, the specific heats of the gas c_p and c_v, and the specific heat of the solid particles, c_{sp}. Similar to Reference [1], we may rewrite the energy equation (3) in the following form:

$$\varrho\left(\frac{\partial}{\partial t} + u\frac{\partial}{\partial y}\right)\left[\frac{p}{\varrho}(1 - Z)\right] + (\Gamma - 1)\,p\frac{\partial u}{\partial y} = 0 \tag{5}$$

where

$$\Gamma = \frac{c_{pm}}{c_{vm}} = \gamma\frac{1 + \delta\beta/\gamma}{1 + \delta\beta} \tag{6}$$

is the ratio of specific heats of the mixture, $\gamma = c_p/c_v$ is the ratio of the specific heats of the gas and $\delta = k_p/(1 - k_p)$ and $\beta = c_{sp}/c_v$.

The boundary conditions at the strong shock, with $u_1 \approx 0$ and $p_1 \approx 0$ are

$$\varrho_1 D = \varrho_2(D - u_2)\,, \qquad p_2 = \varrho_2 u_2(D - u_2)\,, \tag{7a), (7b}$$

$$p_2(D - u_2)\left[\frac{1}{\Gamma - 1}\frac{p_2}{\varrho_2} + \frac{1}{2}u_2^2\right] - p_2 u_2 - \frac{Z_2(D - u_2)\,p_2}{(\Gamma - 1)} = 0 \tag{7c}$$

where subscript 1 refers to the values immediately in front of the strong shock and subscript 2 refers to the values immediately behind the strong shock.

Now we transform the fundamental equations (1), (2) and (5) and boundary conditions (7) into non-dimensional form by introducing the following non-dimensional quantities

$$\eta = y/y_2(t) \tag{8}$$

where y_2 is the shock front at time t and the speed of the shock wave $= D(t) = \mathrm{d}y_2/\mathrm{d}t$. For similarity solution [1], as well as the present non-similar solutions with gravitational forces we have the power-law for the shockwave velocity and is defined as follows:

$$D(t) = D_0(t) = \frac{\mathrm{d}y_2}{\mathrm{d}t} = \frac{\delta y_2}{t}\,. \tag{9}$$

It should be noticed that the relation (9) is not valid for non-similar solution in general.

We now define the non-dimensional variables as follows:

$$i(\eta, t) = \frac{u}{D_0(t)}\,, \qquad j(\eta, t) = \frac{\varrho}{\varrho_0}\,, \qquad k(\eta, t) = \frac{p}{\varrho_0 D_0^2} \tag{10}$$

where $\varrho_1 = \varrho_0$ is the uniform density in front of the shock wave.

We consider a strong shock with a large FROUDE number $Fr = D_0^2/gy_2 = 1/\varepsilon(t)$. Hence we may expand our non-dimensional variables $i(\eta, t)$, $j(\eta, t)$ and $k(\eta, t)$ of equation (10) in terms of powers of $\varepsilon(t) = 1/Fr$. Thus

$$Q(\eta, t) = Q_0(\eta) + \varepsilon(t)\,Q_1(\eta) + O(\varepsilon^2) \tag{11}$$

where $Q(\eta, t)$ may be any one of the non-dimensional functions $i(\eta, t)$, $j(\eta, t)$ and $k(\eta, t)$. The fundamental equations of our problem in terms of η and t are

(i) *Equation of continuity*

$$\frac{y_2}{D_0}\frac{\partial j}{\partial t} - \eta\frac{\partial j}{\partial \eta} + \frac{\partial ij}{\partial \eta} = 0\,, \tag{12}$$

(ii) *Equation of motion*

$$\frac{y_2}{D_0}j\frac{\partial i}{\partial t} + j(i - \eta)\frac{\partial i}{\partial \eta} + \frac{\partial k}{\partial \eta} = \frac{1}{2}ij + \delta_1\varepsilon(t)\,j\,, \tag{13}$$

(iii) *Equation of energy*

$$\frac{y_2}{D_0}\left\{j\frac{\partial}{\partial t}\left[\frac{k}{j}(1 - Z_0 j)\right] + \frac{k}{j}\frac{\partial j}{\partial t}\right\} + \Gamma k\frac{\partial i}{\partial \eta} + (i - \eta)(1 - Z_0 j)\frac{\partial k}{\partial \eta} = k(1 - Z_0 j) \tag{14}$$

where Z_0 is the initial volume fraction of the solid particles in the mixture and Γ is the ratio of specific heat of the mixture.

If we substitute the expression (11) into Eqs. (12) to (14), we find that the zeroth-order solutions $i_0(\eta)$, $j_0(\eta)$ and $k_0(\eta)$ are exactly the same similarity solutions discussed in Reference [1]. The first-order solutions $i_1(\eta)$, $j_1(\eta)$, $k_1(\eta)$ satisfy the following equations:

$$j_0' i_1 (i_0' + 2)\, j_1 + j_0 i_1' + (i_0 - \eta)\, j_1' = 0\,, \tag{15}$$

$$(i_0' + 2)\, j_0 i_1 + (i_0 - \eta)\,(j_1 i_0' + j_0 i_1') + k_1' = \tfrac{1}{2}\,(i_0 j_1 + j_0 i_1) + \delta_1 j_0\,, \tag{16}$$

$$k_1(1 - Z_0 j_0) + \Gamma(k_0 i_1' + i_0' k_1) + (i_0 - \eta)\,[k_1'(1 - Z_0 j_0) - k_0' Z_0 j_1] + i_1 k_0'(1 - Z_0 j_0) = -k_0 Z_0 j_1 \tag{17}$$

where the prime denotes the differentiation with respect to η.

The boundary conditions immediately behind the shock front are

$$i_1(1) = j_1(1) = k_1(1) = 0\,. \tag{18}$$

3. Numerical Solutions and Discussion

The governing equations for the first-order corrections (15) to (17) have been numerically integrated simultaneously with the self-similar equations of Reference [1] using a fourth-order Runge Kutta method subject to the boundary condition (18). The computations were carried out in the high speed computer UNIVAC 1100/40 at the Computer Science Center of the University of Maryland. The integration began at the shock front $\eta = 1$ and terminates at $\eta = 0$, which corresponds to the center of symmetry.

Some of the typical results are presented below.

In Fig. 1 to 3 the effect of variation of k_p and G for a given $\gamma = 1.4$ is presented. In Fig. 1, the velocity correction $i_1(\eta)$ is negative when $\delta_1 = 1.0$ and is positive when $\delta_1 = -1.0$. Increase in k_p decreases the correction $i_1(\eta)$. When G is increased for a fixed k_p and γ we see that the magnitude of velocity correction is larger (negatively) behind the shock but is smaller at $\eta = 0$. Since the gravitational force is uni-directional, the velocity symmetry condition $i(0) = 0$ is not satisfied and $i_1(\eta)$ has a finite value at $\eta = 0$. In Fig. 2, the density correction $j_1(\eta)$ for various k_p and G are shown. The density correction is negative behind the shock and positive near the center when $\delta_1 = 1.0$, and vice versa when $\delta_1 = -1.0$. Increase in k_p for fixed G and γ shows that the peak correction shifts closer to the center. There is no density correction at $\eta = 0$; i.e., $j_1(0) = 0$. Increase in G for fixed γ and k_p decreases the magnitude of the density correction overall and the maximum values are lower. In Fig. 3, the pressure correction $k_1(\eta)$ for various k_p and G are shown. The solutions indicate a pressure "plateau" as the center is approached. As k_p is increased, the correction decreases throughout the flow. However, when G is increased, the magnitude of pressure correction increases (negatively, when $\delta_1 = 1.0$) behind the shock but decreases as the center ($\eta = 0$) is approached.

In Fig. 4 to 6, the effect of variation of γ for a given k_p and G is shown. In Fig. 6, the velocity correction $i_1(\eta)$ is shown to increase continuously as the center is approached and as γ is increased, the correction $i_1(\eta)$ also increases. In Fig. 5, the density correction $j_1(\eta)$ is shown. Increase in γ continuously shifts the peak value towards the center ($\eta = 0$). The value of $j_1(\eta) \to 0$ as $\eta \to 0$. Fig. 6 shows the variation of $k_1(\eta)$ with variation in γ. Increase in γ increases the value of $k_1(\eta)$ throughout the flow. Figs. 7 to 9 show the effect of the gravitational correction as compared to the self-similar case [1] for fixed ε. In Fig. 7, the velocity ratio u/u_2 behind the shock wave is shown to be less when the gravitational effect is included for all combinations of γ, G and k_p. In Fig. 8, the total density ratio ϱ/ϱ_2 behind the shock wave as a function of similarity parameter is shown. The density ratio decreases behind the shock, when gravitational force is included, and increases as the center is approached in comparison to the self-similar case. The variation of pressure ratio behind the shock with similarity parameter η is shown in Fig. 9. The pressure is consistently lower when the gravitational force is included (when $\delta_1 = 1.0$). It must be mentioned here that all the trends shown here reverse when the direction of gravitational force is reversed. In Fig. 10, we see that the shock velocity $D(t)$ is faster when gravitational force (with $\delta_1 = 1.0$) is included. The velocity also increases with increase in γ and E_0. Here E_0 is the energy release associated with the strong shock and is given by

$$E_0 = 2 \int_0^{y_2} \varrho \left(\frac{1 - Z}{\Gamma - 1} \frac{p}{\varrho} + \frac{1}{2} u^2 \right) \mathrm{d}y\,. \tag{19}$$

By dimensional analysis [2], we may write

$$D(t) = \frac{\mathrm{d}y_2}{\mathrm{d}t} = \frac{2}{3} \frac{y_2}{t} = \frac{2}{3} \left(\frac{E_0}{\varrho_0 \alpha} \right)^{1/2} y_2^{-1/2} = \frac{2}{3} \left(\frac{E_0}{\alpha \varrho_0} \right)^{1/3} t^{-1/3} \tag{20}$$

and we have

$$\alpha = \frac{8}{9} \int_0^1 \left[\frac{(1 - Z_0 j)}{(\Gamma - 1)}\, k + \frac{1}{2}\, j i^2 \right] \mathrm{d}\eta\,, \tag{21}$$

α is determined using the values with gravitational correction included for an assumed small value for $\varepsilon(t)$.

Fig. 11 shows the variation of shock front location $y_2(t)$ as a function t and γ. As the shock is traveling faster when gravitational force is included the shock location is also increased. This increase is faster when γ is large. And finally, in Fig. 12, the variation of the expansion parameter $\varepsilon(t) = 1/Fr$ is shown as a function of time t. This figure

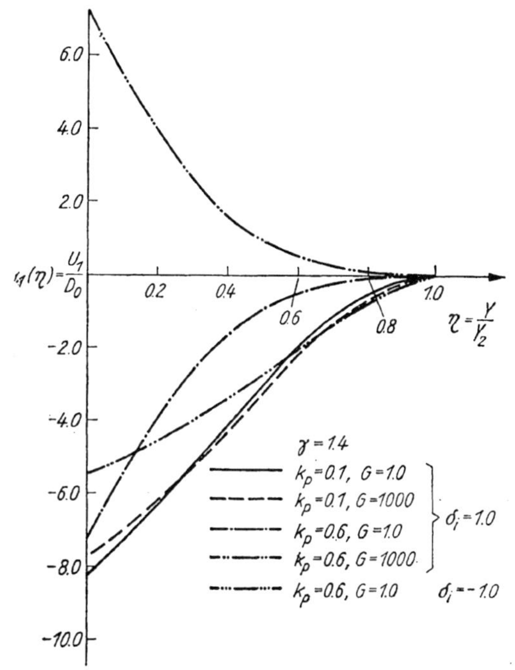

Fig. 1. The variation of the first-order velocity correction $i_1(\eta)$ at $\gamma = 1.4$ for various k_p and G

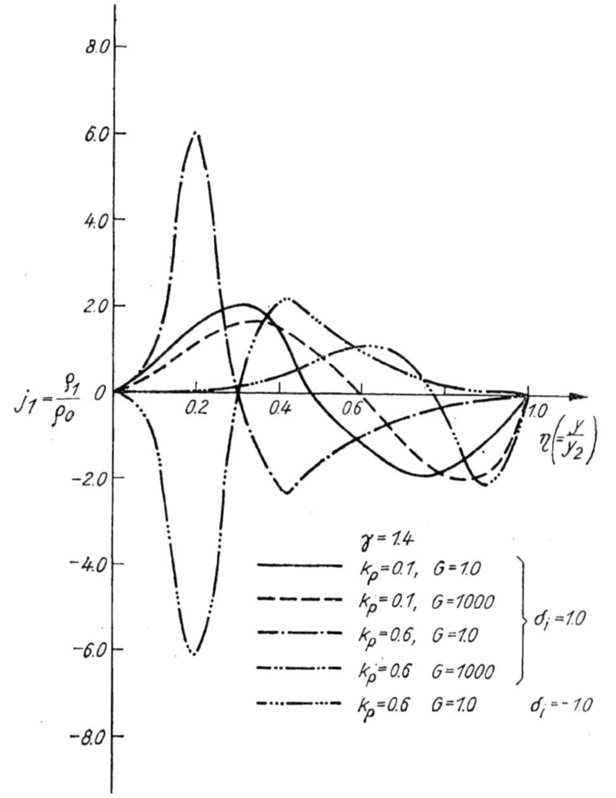

Fig. 2. The variation of the first-order density correction $j_1(\eta)$ at $\gamma = 1.4$ for various k_p and G.

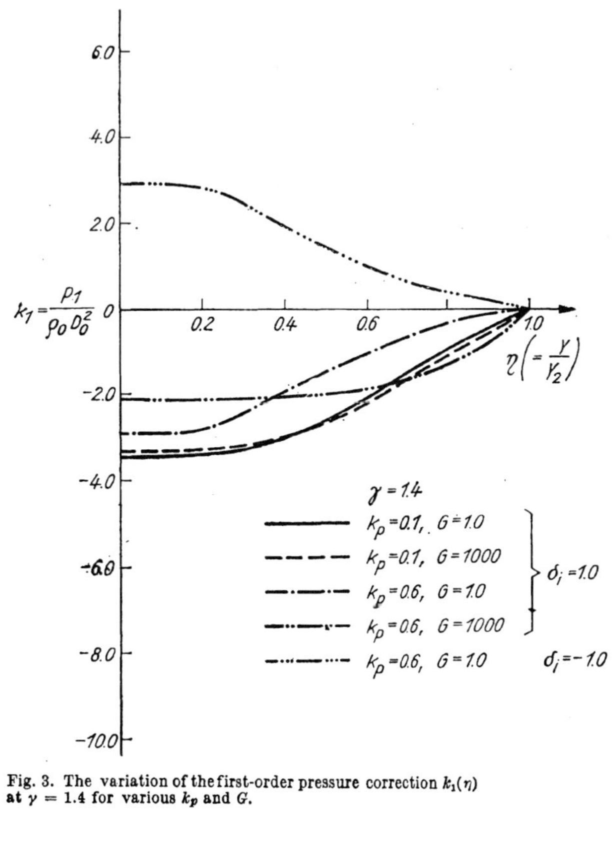

Fig. 3. The variation of the first-order pressure correction $k_1(\eta)$ at $\gamma = 1.4$ for various k_p and G.

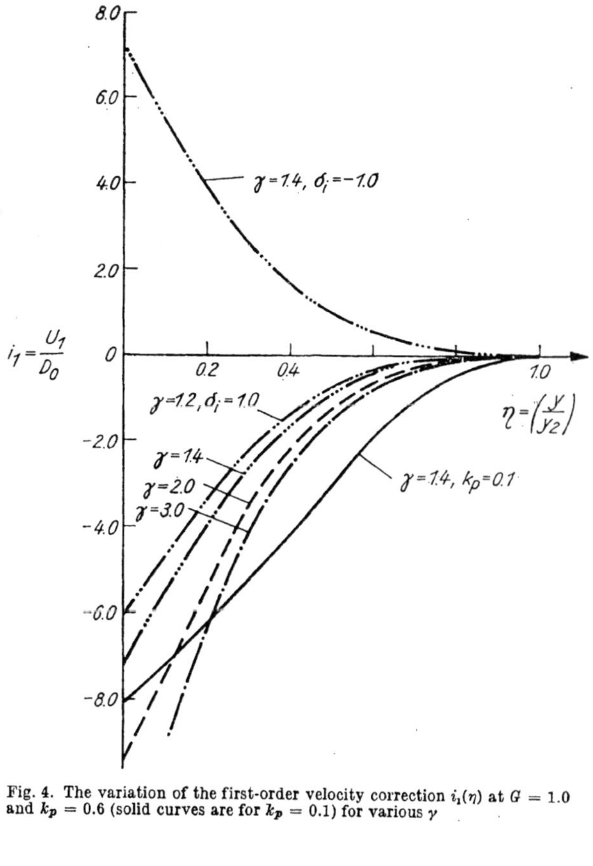

Fig. 4. The variation of the first-order velocity correction $i_1(\eta)$ at $G = 1.0$ and $k_p = 0.6$ (solid curves are for $k_p = 0.1$) for various γ

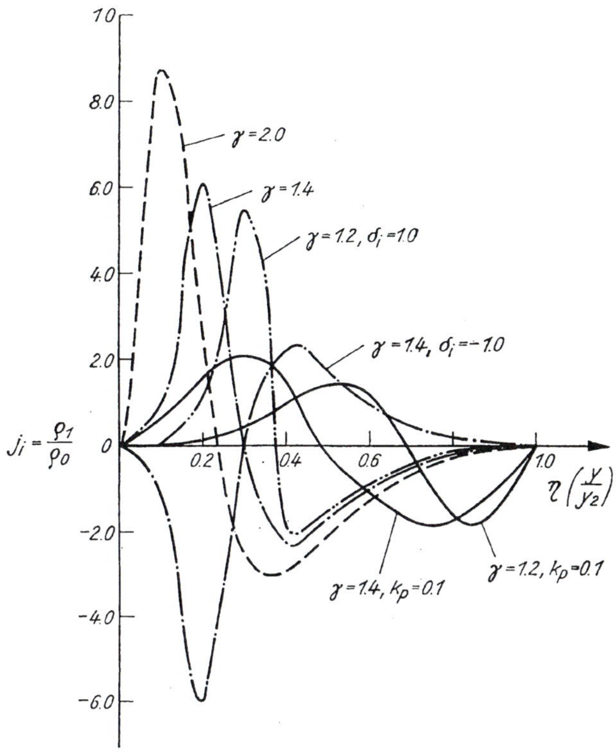

Fig. 5. The variation of the first-order density correction $j_1(\eta)$ at $G = 1.0$ and $k_p = 0.6$ (solid curves are for $k_p = 0.1$) for various γ

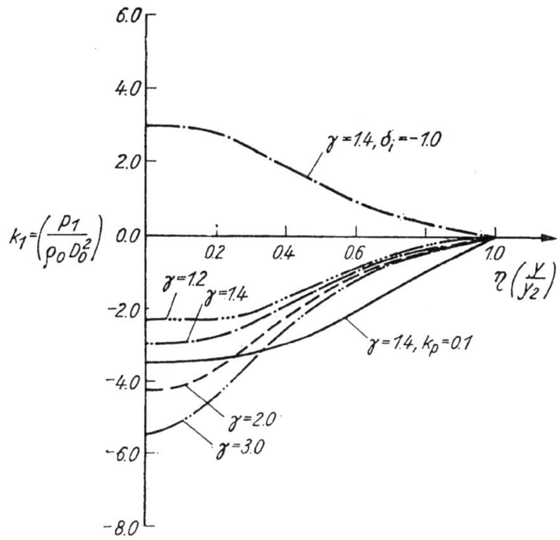

Fig. 6. The variation of the first-order pressure correction $k_1(\eta)$ at $G = 1.0$ (solid curves are for $k_p = 0.1$) for various γ

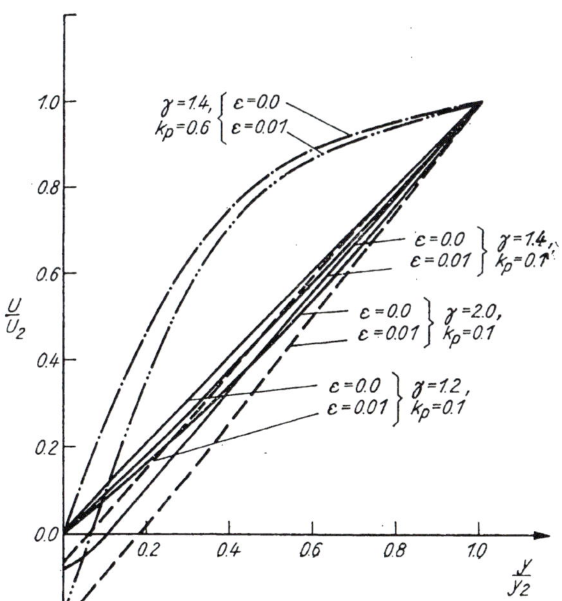

Fig. 7. The variation of the velocity ratio (u/u_2) behind the shock wave with the similarity parameter $\eta\,(= y/y_2)$ at $G = 1.0$ for various γ and k_p

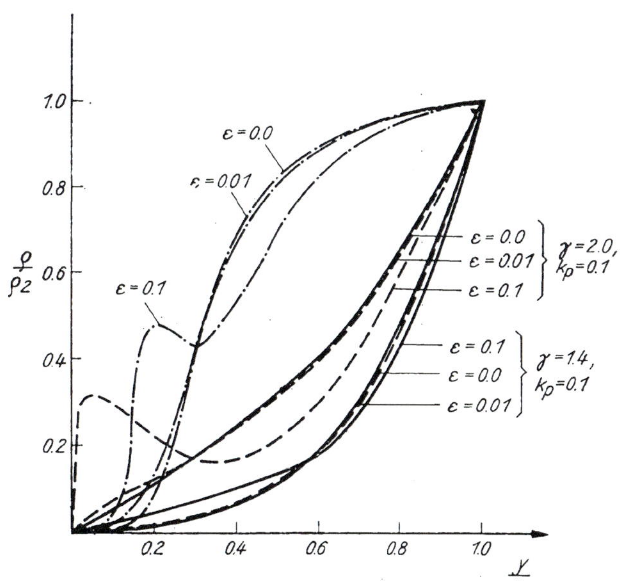

Fig. 8. The variation of the density ratio (ρ/ρ_2) behind the shock wave with the similarity parameter $\eta\,(= y/y_2)$ at $G = 1.0$ for various γ and k_p

indicates that for the small time, $\varepsilon(t)$ is very small and so this linearization is valid for the strong shock approximations.

In conclusion, we see that the inclusion of the gravitational force leads to some fundamental changes in the pressure, density and velocity histories behind the propagating strong shock. The effect is highly depended on the non-dimensional parameters k_p and G [1], indicating the complexity of a two-phase analysis.

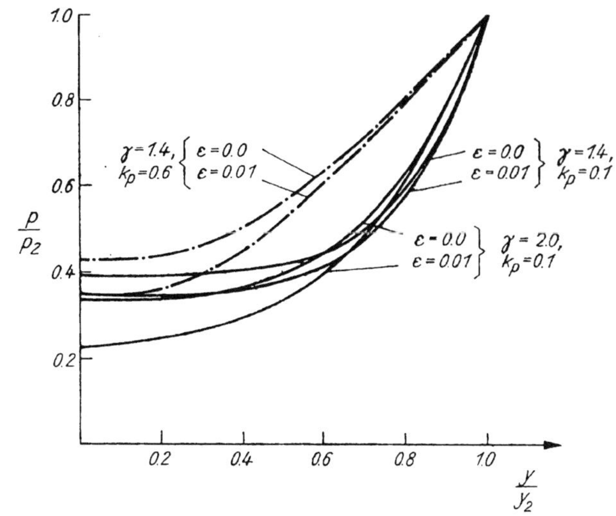

Fig. 9. The variation of the pressure ratio (p/p_2) behind the shock wave with the similarity parameter $\eta (= y/y_2)$ at $G = 1.0$ for various γ and k_p

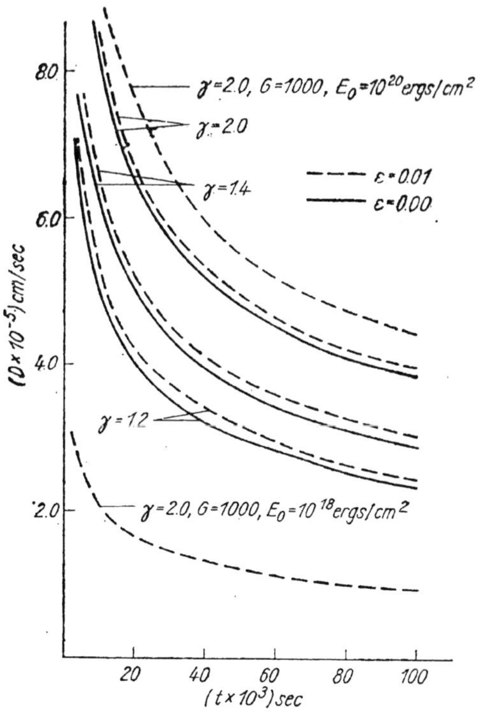

Fig. 10. The variation of the shock velocity D as a function of time at $k_p = 0.6$, $G = 1.0$ and $E_0 = 10^{18}$ ergs/cm² (unless otherwise indicated) for various γ

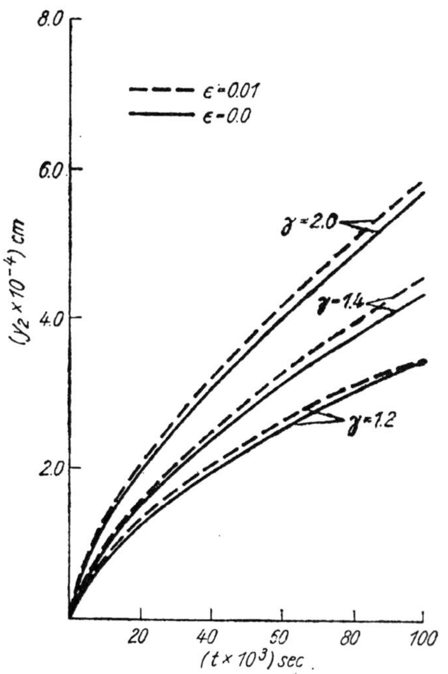

Fig. 11. The variation of the shock location y_2 as a function of time at $k_p = 0.6$, $E_0 = 10^{18}$ ergs/cm² and $G = 1.0$ for various γ

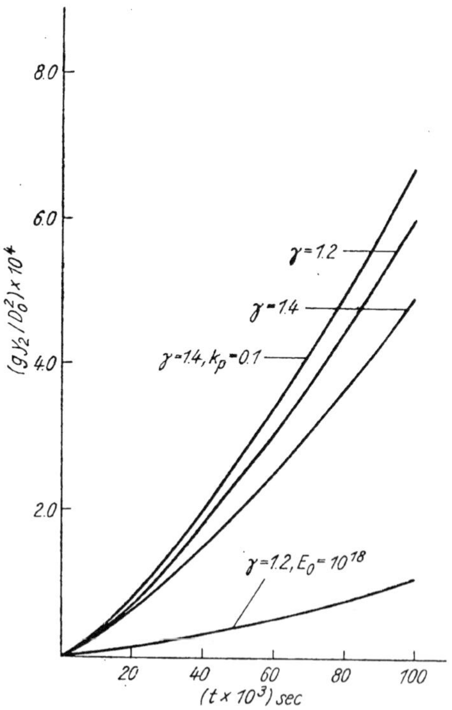

Fig. 12. The variation of the perturbation parameter $\varepsilon = g y_2/D_0^2$ as a function of time at $G = 1.0$, $k_p = 0.6$ and $E_0 = 10^{18}$ erg/cm² (unless otherwise indicated) for various γ

References

1 PAI, S. I., MENON, S., FAN, Z. Q., Similarity Solutions of a Strong Shock Wave Propagation in a Mixture of a Gas and Dusty Particles, to be published in the International Journal of Engineering Science, 1980.

2 SCHULTZ, P. H., SRNKA, L. J., PAI, S. I., and MENON, S., Cometary Collisions on the Moon and Mercury, paper presented at the XIth Lunar and Planetary Conference, Houston, Texas, March 17–21, 1980.

Eingereicht am 30. 7. 1980

Anschrift: Prof. Dr. S. I. PAI, Institut für Strömungslehre und Strömungsmaschinen, Universität Karlsruhe, D-7500 Karlsruhe, Kaiserstraße 12, BRD

ZAMM 61, 215–220 (1981)

Goodarz Ahmadi

Squeeze Film Theory for Cosserat Fluids

Es wird die Schmierfilmtheorie für Flüssigkeiten mit innerer Struktur behandelt. Ausgehend von den allgemeinen Bewegungsgleichungen für ein Cosserat-Fluid wird der Fall der schleichenden Parallelströmung untersucht. Für die Geschwindigkeitsverteilung in einem engen Spalt wird eine Formel hergeleitet. Die Abhängigkeit der Tragkraft des Films von der Filmdicke und einer charakteristischen Länge der inneren Struktur wird berechnet, außerdem wird eine Zeit-Höhe-Relation hergeleitet und diskutiert. Es werden die Grenzfälle sehr kleiner und sehr großer charakteristischer Länge der inneren Struktur betrachtet: Für die Tragkraft und die Zeit-Höhe-Relation werden einfache Ausdrücke gewonnen. Die Ergebnisse werden in mehreren Abbildungen graphisch dargestellt. Schließlich wird die charakteristische Länge der inneren Struktur des Cosserat-Fluids mit Hilfe der experimentellen Daten für Schmierfilme abgeschätzt.

The fundamental theory of squeeze film of fluids with inner structure is considered. The basic equations of motion of Cosserat fluids are reviewed. The special case of creeping parallel flow is investigated and the expression for the velocity distribution in a narrow channel is derived. The variation of film bearing capacity with film thickness and inner structure length scale is calculated and the expression for time-height relationship is also obtained and discussed. The limits of small and large fluid inner structure length measure are considered and simple expressions for the load supporting capacity and time-height relationship are derived. The results are plotted in several figures. The inner structure of the Cosserat fluid is also estimated from the existing experimental data for thin squeeze films.

Рассматривается теория смазочной пленки для жидкостей с внутренной структурой. Исходя из общих уравнений движения жидкости Коссерата исследован специальный случай ползучего параллельного движения. Найден выражение для разпределения скорости в узкой щели. Зависимость подьемной силы пленки от толщиной пленки и от характерной длины внутренной структуры вычисляется и кроме того получен зависимость высоты от времени, которые обсуждены. Рассматриваются предельные случая очень маленьких и очень больших характерных длин внутренной структуры. Для подьемной силы и для зависимости высоты от времени найдены простые выражения. Результаты приведены в несколько графиков. Характерная длина внутренной структуры жидкости Коссерата оценивается с помощью экспериментальных данных для смазочной пленки.

1. Introduction

The possibility of existence of nonvanishing couple stresses has been first conceived by E. and F. Cosserat [1]. The modern developments of the theory of media with micro-structure are due to Truesdell and Toupin [2], Toupin [3], Mindlin and Tierstin [4], Eringen and Suhubi [5], Eringen [6, 7], and Mindlin [8]. The mechanics of fluids with couple stress have been studied by Grad [9], Ericksen [10, 11], Eringen [6, 12], Condiff and Dahler [13], Kline and Allen [12], Bleustein and Green [15], Stokes [16], Shahinpoor and Ahmadi [17], Ahmadi, Koh, and Goldschmidt [18, 19], and Ahmadi [20, 21] among others. Recently, the theory of lubrication of fluids with microstructure have been considered by Allen and Kline [22], Balaram and Sastri [23], and Ahmadi [24]. Agrawal, Ganju and Jethi [25], and Prakash and Sinha [26] have investigated the theory of squeeze film of a micropolar fluid.

In the present work the theory of squeeze film of a Cosserat fluid is studied. The basic equations of balance and corresponding constitutive equations are first reviewed. The problem of creeping parallel flow is studied and the expression for the velocity distribution in a narrow channel is obtained and discussed. The expressions for pressure variation, load bearing capacity and time-height relationship as functions of film thickness and inner structure length measure are derived. The limits of small and large fluid inner structure length measure are considered and simplified expressions for load bearing capacity and time are developed. Both cases of rectangular and circular squeeze plates are treated. It is observed that the load bearing capacity and the response time are increased with an increase in the length microscale. The results are presented in several elaborated graphs. Finally the theoretical predictions are compared with the experimental findings of Needs [27] and the corresponding value of the inner structure length measure is estimated which is shown to be of the order of molecular length scale.

2. Basic Equations

The basic laws of motion of a Cosserat fluid as derived in [15–17] are

$$\text{Mass:} \quad \frac{\partial \varrho}{\partial t} + \nabla \cdot (\varrho \boldsymbol{v}) = 0\,, \tag{1}$$

$$\text{Linear Momentum:} \quad \nabla \cdot \boldsymbol{\tau} + \varrho \boldsymbol{f} = \varrho \dot{\boldsymbol{v}}\,, \tag{2}$$

$$\text{Angular Momentum:} \quad \nabla \cdot \boldsymbol{\mu} + \boldsymbol{I} \times \boldsymbol{\tau} + \varrho \boldsymbol{c} = 0\,, \tag{3}$$

Energy:

$$\varrho \dot{e} = \boldsymbol{\tau} : \nabla \boldsymbol{v} - \tfrac{1}{2} (\boldsymbol{\tau} \times \boldsymbol{I}) \cdot \nabla \times \boldsymbol{v} + \tfrac{1}{2} \boldsymbol{\mu} : \nabla\nabla \times \boldsymbol{v} + \nabla \cdot \boldsymbol{q} \tag{4}$$

where ϱ is the mass density, $\boldsymbol{v}$ is the velocity vector, $\boldsymbol{\tau}$ is the stress tensor, $\boldsymbol{\mu}$ is the couple stress tensor, $\boldsymbol{f}$ is the body force per unit mass, $\boldsymbol{c}$ is the body couple per unit mass, e is the internal energy density per unit mass and $\boldsymbol{q}$ is the heat flux vector pointing outward.

Noting that

$$\boldsymbol{\tau} = \boldsymbol{\tau}^s + \boldsymbol{\tau}^a \tag{5}$$

where $\boldsymbol{\tau}^s$ and $\boldsymbol{\tau}^a$ are the symmetric and antisymmetric parts of the stress tensor and evaluating $\boldsymbol{\tau}^a$ from equation (3) and substituting into equation (2), the result is

$$\nabla \cdot \boldsymbol{\tau}^s + \tfrac{1}{2} \nabla \times \nabla \cdot \boldsymbol{\mu}^d + \varrho \boldsymbol{f} + \tfrac{1}{2} \nabla \times \varrho \boldsymbol{c} = \varrho \dot{\boldsymbol{v}}\,, \tag{6}$$

where $\boldsymbol{\mu}^d$ is the devitoric part of the couple stress tensor. The following set of linear constitutive equations have been proposed [15—17] for a COSSERAT fluid

$$\boldsymbol{\tau}^s = (-p + \lambda \nabla \cdot \boldsymbol{v})\, I + \mu(\nabla \boldsymbol{v} + \boldsymbol{v}\nabla)\,, \tag{7}$$

$$\boldsymbol{\mu}^d = 2\eta \nabla\nabla \times \boldsymbol{v} + 2\eta' \nabla \times \boldsymbol{v}\nabla\,, \tag{8}$$

where p is the hydrostatic pressure and λ, μ, η and η' are coefficients of viscosity restricted to the following inequalities:

$$\mu \geqq 0\,, \qquad 3\lambda + 2\mu \geqq 0\,, \qquad \eta \geqq 0\,, \qquad |\eta'| \leqq \eta\,. \tag{9}$$

Based on (6—8) and in the absence of body force and body couple, the equation of motion becomes

$$\varrho \dot{\boldsymbol{v}} = -\nabla p + \mu \nabla^2 \boldsymbol{v} + (\lambda + \mu)\, \nabla\nabla \cdot \boldsymbol{v} + \eta \nabla^2 \nabla \times \nabla \times \boldsymbol{v}\,. \tag{10}$$

It has been shown in [15, 25] that for sufficiently low REYNOLDS number and when the velocity and the tangential component of angular velocity are specified on the surface boundary, the solution of equation (10) is stable and unique.

This completes our summary of basic equations of a COSSERAT fluid.

3. Squeeze Films Theory

In this section the creeping plane parallel flow of a COSSERAT fluid is first reviewed. Assuming that

$$\boldsymbol{v} = \big(u(y), 0, 0\big)\,, \qquad p = p(x)\,, \tag{11}$$

the equation of motion (10) reduces to

$$l^2 \frac{\partial^4 u}{\partial y^4} - \frac{\partial^2 u}{\partial y^2} = -\frac{1}{\mu} \frac{\partial p}{\partial x}\,, \qquad 0 < y < h(x)\,, \tag{12}$$

where

$$l^2 = \eta/\mu \tag{13}$$

and $h(x)$ is the width of the gap and the parameter l which has a dimension of length is associated with the size of the internal structure of the fluid. For boundary conditions it is assumed that at the surface boundaries the velocity is given and the couple stress vanishes i.e.

$$u = U\,, \quad \frac{\partial^2 u}{\partial y^2} = 0 \quad \text{at} \quad y = 0; \qquad u = 0\,, \quad \frac{\partial^2 u}{\partial y^2} = 0 \quad \text{at} \quad y = h\,. \tag{14}$$

In other words it is assumed that the surface boundaries are nonpolar and hence incapable of supporting couple stresses.

It has been shown in [24] that the velocity distribution satisfying equation (12) and boundary conditions (14) is given by

$$u = \frac{l^2}{\mu} \frac{\partial p}{\partial x} [1 - \cosh (y/l)] + \frac{h^2}{2\mu} \frac{\partial p}{\partial x} \left\{ -(y/h) + (y/h)^2 + 2(l/h)^2 [\cosh (h/l) - 1] \frac{\sinh (y/l)}{\sinh (h/l)} \right\} + U(1 - y/h)\,. \tag{15}$$

The volume flow rate per unit width is given by

$$q = \int_0^h u \, \mathrm{d}y\,. \tag{16}$$

Evaluating the integral, it is found that [24]

$$q = \frac{Uh}{2} + \frac{h^3}{\mu} \frac{\partial p}{\partial x} \left[-\frac{1}{12} + (l/h)^2 - 2(l/h)^3 \tanh (h/2l) \right]. \tag{17}$$

Now consider the problem of squeeze film of a COSSERAT fluid as shown schematically in figure 1. The upper plate moves with a squeeze velocity V_0 toward the stationary lower plate. Order of magnitude analysis shows that for a sufficiently small gap the assumption of creeping parallel flow between the plates is reasonably accurate.

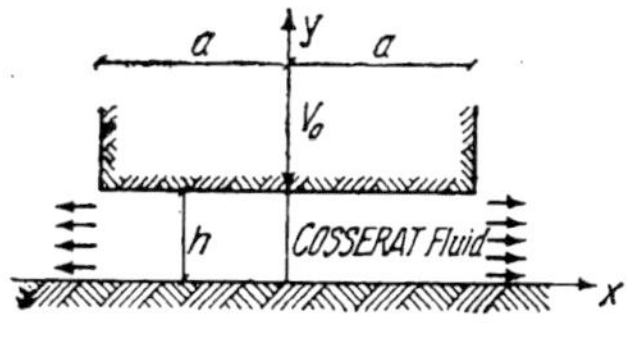

Fig. 1. Schematic configuration of squeeze film model

The equation of continuity for the film becomes

$$\frac{\partial q}{\partial x} = V_0 . \tag{18}$$

Employing expression (17) into equation (18) gives

$$\frac{d^2 p}{dx^2} = -\frac{\mu V_0}{h^3 f(h, l)}, \tag{19}$$

where

$$f(h, l) = \frac{1}{12} - \left(\frac{l}{h}\right)^2 + 2\left(\frac{l}{h}\right)^3 \tanh\left(\frac{h}{2l}\right). \tag{20}$$

Equation (19) gives the differential equation for evaluation of film pressure for a rectangular plate. Similarly, use of continuity equation in polar coordinates yields

$$\frac{1}{r}\frac{d}{dr}\left(r\frac{dp}{dr}\right) = -\frac{\mu V_0}{h^3 f(h/l)}, \tag{21}$$

for circular plates.

i) Analysis for Rectangular Plates

The pressure distribution for squeeze films between two rectangular plates of length $2a$ is found by integration of equation (19), i.e.

$$p - p_0 = \frac{\mu V_0 (a^2 - x^2)}{2h^3 f(h, l)}, \tag{22}$$

where the following boundary conditions are employed:

$$p = p_0 \quad \text{at} \quad x = \pm a . \tag{23}$$

The instantaneous load supporting capacity of the squeeze film is given by

$$W = b \int_{-a}^{a} (p - p_0)\, dx = \frac{2\mu V_0 b a^3}{3h^3 f(h, l)}, \tag{24}$$

where b is the width of the plates.

Introducing dimensionless quantities

$$H = h/h_0 , \qquad \varepsilon = l/h_0 , \tag{25}$$

where h_0 is the initial film thickness, the nondimensionalized load bearing capacity becomes

$$\overline{W} = \frac{3h_0^2 W}{2\mu V_0 b a^3} = \frac{1}{H^3 F(H, \varepsilon)}, \tag{26}$$

where

$$F(H, \varepsilon) = \frac{H^3}{12} - \varepsilon^2 H + 2\varepsilon^3 \tanh\left(\frac{H}{2\varepsilon}\right). \tag{27}$$

Assuming the load to be constant and noting that

$$V_0 = \frac{dh}{dt}, \tag{28}$$

equation (24) becomes a differential equation for film thickness as a function of time. The time t required for reducing the height h_0 to a prescribed film thickness h then is given by

$$T = \frac{W h_0^2 t}{4\mu b a^3} = \frac{1}{6}\int_{H}^{1} \frac{dH}{F(H, \varepsilon)}, \tag{29}$$

where T is the nondimensionalized time.

In general the integral in equation (29) must be evaluated numerically. However, in the limiting cases of small or large ε, it is possible to find approximate analytical expressions for the dimensionless time. For the case of small ε, that is when the length measure of the internal structure of the fluid is much smaller than the film thickness, equation (27) may be approximated by

$$F(H, \varepsilon) = \frac{H^3}{12} - \varepsilon^2 H , \quad \text{as} \quad \varepsilon \to 0 . \tag{30}$$

Employing equation (30) into expression (29) and carrying out the required integration, it follows that

$$T = H^{-2} - 1 + 6\varepsilon^2 (H^{-4} - 1) , \quad \text{as} \quad \varepsilon \to 0 . \tag{31}$$

Equation (31) gives the approximate expression for the dimensionless time for the limit of small ε. Clearly when $\varepsilon = 0$, equation (31) reduces to the results of a NEWTONian fluid.

When the inner length measure of the COSSERAT fluid is much larger than the film thickness, that is, the limit of large ε, equation (27) may be approximated by

$$F(H, \varepsilon) = \frac{H^5}{120\varepsilon^2} - \frac{17H^7}{20160\varepsilon^4}, \quad \text{as} \quad \varepsilon \to \infty . \tag{32}$$

The corresponding expression for the dimensionless time then becomes

$$T = 5\varepsilon^2(H^{-4} - 1) + \tfrac{85}{84}(H^{-2} - 1), \quad \text{as} \quad \varepsilon \to \infty . \tag{33}$$

From equations (31) and (33) it is concluded that a COSSERAT fluid film has larger load bearing capacity in comparison to a NEWTONian fluid with the same coefficient of viscosity and therefore the time to reduce the film thickness is larger for a COSSERAT fluid. Similar observations are also made in [24—26].

ii) Analysis for Circular Plates

The pressure variation for squeeze film of a COSSERAT fluid between two circular plates of radius a is found by integration of equation (21), i.e.

$$p - p_0 = \frac{\mu V_0(a^2 - r^2)}{4h^3 f(h, l)}, \tag{34}$$

where the boundary conditions

$$p = p_0 \quad \text{at} \quad r = a , \qquad \frac{dp}{dr} = 0 \quad \text{at} \quad r = 0, \tag{35}$$

have been employed. The instantaneous load bearing capacity of the film then is given by

$$W = \int_0^a 2\pi r(p - p_0)\, dr = \frac{\pi\mu V_0 a^4}{8h^3 f(h, l)}. \tag{36}$$

Employing the nondimensional quantities of equation (25), the expression for the nondimensionalized load bearing capacity becomes

$$\overline{W} = \frac{8Wh_0}{\pi\mu V_0 a^4} = \frac{1}{H^3 F(H, \varepsilon)}, \tag{37}$$

where $F(H, \varepsilon)$ is given by equation (27).

Similar to the case of rectangular plates, time-height relationship may be estabilished for the present case of circular plates, i.e.

$$T = \frac{4Wh_0^2 t}{3\pi\mu a^4} = \frac{1}{6}\int_H^1 \frac{dH}{H^3 F(H, \varepsilon)}. \tag{38}$$

By comparison of equations (37) and (38) with equations (26) and (29) it follows that the expressions for dimensionless load bearing capacity $\overline{W}$ and dimensionless time T for the circular plates are identical to those of the rectangular case. Therefore, the results presented in equations (30)—(33) for the limiting analysis of small and large ε are also valid for the case of circular plates.

4. Results and Discussions

The variations of dimensionless instantaneous load bearing capacity with film thickness for several values of ε are shown in figure 2. It is observed that the load bearing capacity of the film increases rapidly with a decrease in the film thickness. Furthermore, the load bearing capacity is also a strong function of ε and increases with an increase in ε.

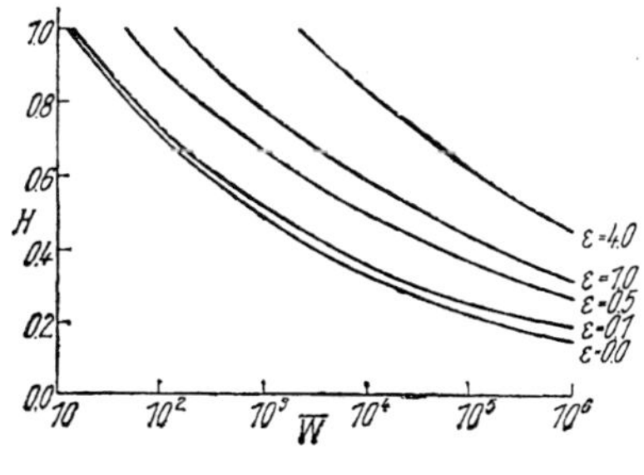

Fig. 2. Variations of dimensionless load with dimensionless film thickness for different values of ε

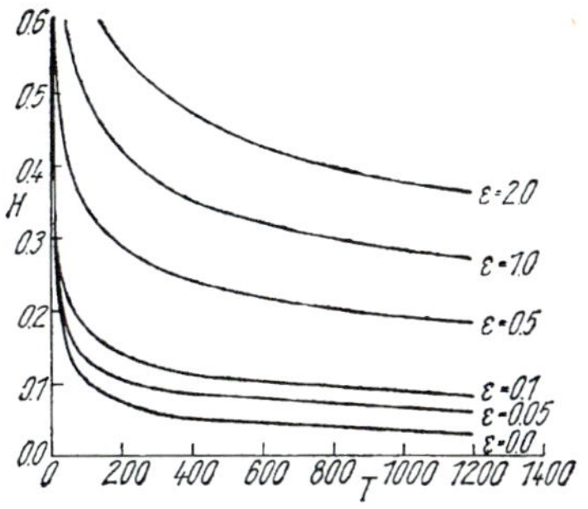

Fig. 3. Variation of dimensionless time with dimensionless film thickness for several values of ε

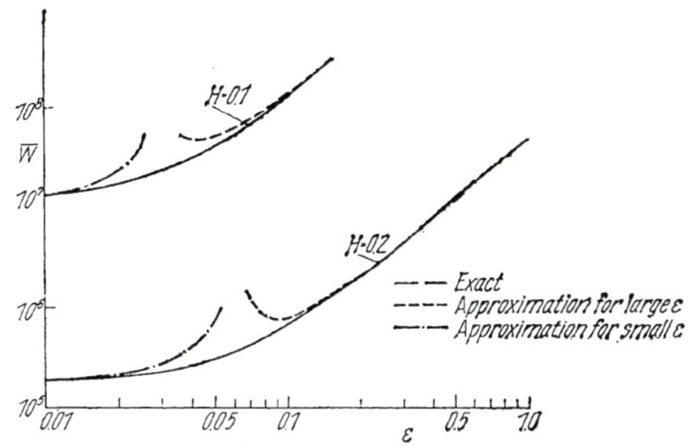

Fig. 4. Variation of dimensionless load bearing capacity with ε for different values of film thickness

Figure 3 shows the variations of nondimensionlized time with dimensionless film thickness for several values of ε. It is observed that the dimensionless time increases with a decrease in the film thickness H. Furthermore, T also increases rapidly with an increase in the length scale of the fluid. For instance, from figure 3 it follows that for the film thickness to reach one tenth of its original size, the time required for a COSSERAT fluid with ε being 0.05 and 0.1 are respectively, 2.2 and 6 times that of a NEWTONian fluid ($\varepsilon = 0$).

Figure 4 shows the log-log plot of the variations of the nondimensionalized load supporting capacity with ε for two values of dimensionless thickness $H = 0.1$ and 0.2. The approximate expressions for the limit of large and small ε with the values of $F(H, \varepsilon)$ given by equations (30) and (32) respectively are also shown in that figure. The relatively steep increase of $\overline{W}$ with ε is clearly seen from figure 4. Furthermore, it is also observed that the approximate expression for the limit of small ε is quite accurate for ε less than 0.02 and the range of accuracy increases with H. The approximate expression for the limit of large ε appears to be remarkably accurate. For $H = 0.1$, it is accurate for $\varepsilon \geqq 0.05$ and for $H = 0.2$, it is accurate for $\varepsilon \geqq 0.1$.

The variations of dimensionless time with nondimensionalized length microscale for different values of dimensionless film thickness are shown in figure 5. It is observed that the dimensionless time increases rapidly with an increase in ε. The predictions of approximate equation (31) for the dimensionless time in the limit of small ε are also shown in figure 5. It is observed that the prediction of equation (31) for H equal to 0.1 and 0.2 are quite accurate for ε less than 0.05 and 0.1, respectively.

In the range of values considered in figure 5, the predictions of approximate expression (33) for the dimensionless time in the limit of large ε are remarkably accurate and almost indistinguishable from the exact solution and hence are not plotted in this figure. Apparently, equation (33) losses its accuracy for values of ε below 0.01.

In [26], it was shown that a polar fluid model can explain the experimental finding by NEEDS [27]. Here, we employ the results of NEED's experiment to estimate the parameter ε and the corresponding inner structure length scale l. NEEDS [27] has used Castor oil with a viscosity of 0.3503 N-sec/m^2 between two circular disks of diameter 19.05 mm. He measured the time for the film thickness to reduce from 0.00406 to 0.00127 mm at a constant load of 0.885 N to be 4680 sec. Theoretical time prediction based on NEWTONian fluid theory gives 4296 sec.

In other words the ratio of the actual measured time to the NEWTONian fluid predicted time T_N is

$$\frac{T}{T_N} = 1.0894 . \tag{39}$$

From equation (31) the ratio of predicted time for COSSERAT fluid to that of a NEWTONian fluid ($\varepsilon = 0$) is found to be

$$\frac{T}{T_N} \approx 1 + 6\varepsilon^2(H^{-2} + 1) . \tag{40}$$

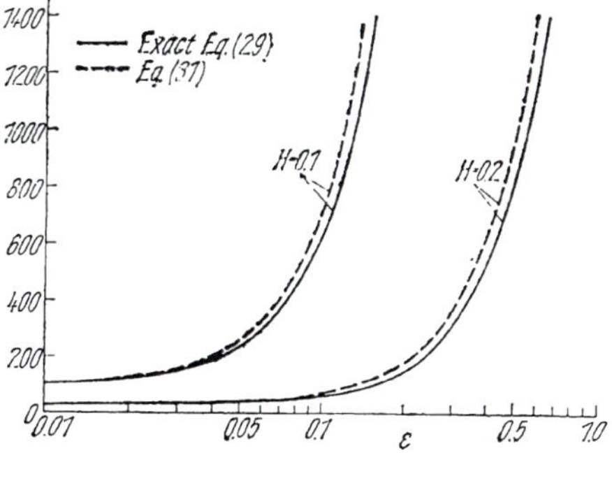

Fig. 5. Variations of dimensionless time with ε for two values of dimensionless film thickness

For the experimental results of NEEDS the value of H was 0.313. The value of ε may then be found from equations (39) and (40), i.e.

$$\varepsilon = 0.036\,. \tag{41}$$

The corresponding values for the inner structure length scale becomes

$$l = 1.48 \times 10^{-4}\ \text{mm}\,. \tag{42}$$

Similar computation based on expression (33) for large values of ε gives

$$\frac{T}{T_N} \approx 5\varepsilon^2(1 + H^{-2}) + 85/84 \tag{43}$$

which yields $\varepsilon = 0.037$ and $l = 1.51 \times 10^{-4}$ mm. Therefore the value of the inner structure length measure is of the order of the molecular length scale.

Therefore, the present COSSERAT fluid model is quite capable of explaning NEED's [27] results of enhancement of response time for very thin squeeze films with some reasonable value of inner structure length measure.

Acknowledgment

The partial financial supports of the Ministry of Education and Higher Education of Iran and the Shiraz University Research Council is gratefully acknowledged.

References

1 COSSERAT, E. and F., Theories des Corps Deformable, Paris (1909).
2 TRUESDELL, C. and TOUPIN, R. A., The Classical Field Theories, Encyclopedia of Physics, Vol. III/1, Ed. S. FLÜGGE, Springer-Verlag, Berlin (1960).
3 TOUPIN, R. A., Elastic materials with couple stress, Arch. Rat. Mech. Anal. **11**, 385 (1962).
4 MINDLIN, R. D. and TIERSTEN, H. F., Effect of couple stresses in linear elasticity, Arch. Rat. Mech. Anal. **11**, 415 (1962).
5 ERINGEN, A. C. and SUHUBI, E. S., Non-linear theory of simple microelastic solids-I, Int. J. Engng Sci. **2**, 189 (1964).
6 ERINGEN, A. C., Simple microfluids, Int. J. Engng Sci. **2**, 205 (1964).
7 ERINGEN, A. C., Linear theory of micropolar elasticity, J. Math. Mech. **15**, 909 (1966).
8 MINDLIN, R. D., Microstructure in linear elasticity, Arch. Rat. Mech. Anal. **16**, 51 (1964).
9 GRAD, H., Statistical mechanics, thermodynamics, and fluid dynamics of systems, Comm. Pure Appl. Math. **5**, 455 (1952).
10 ERICKSEN, J. L., Anisotropic fluids, Arch. Rat. Mech. Anal. **4**, 231 (1960).
11 ERICKSEN, J. L., Transversely isotropic fluids, Kolloidzeitschrift **173**, 117 (1960).
12 ERINGEN, A. C., Theory of micropolar fluids, J. Math. Mech. **16**, 1 (1966).
13 CONDIFF, D. W. and DAHLER, J. S., Fluid mechanical aspects of antisymmetric stress, Phys. Fluids **7**, 842 (1964).
14 KLINE, K. A. and ALLEN, S. J., Nonsteady flows of fluids with microstructure, Physics Fluids **13**, 263 (1970).
15 BLEUSTEIN, J. L. and GREEN, A. E., Dipolar fluids, Int. J. Engng. Sci. **5**, 323 (1967).
16 STOKES, V. K., Couple stresses in fluids, Phys. Fluids **9**, 1709 (1966).
17 SHAHINPOOR, M. and AHMADI, G., Stability of Cosserat fluid motions, Arch. Rat. Mech. Anal. **47**, 183 (1972).
18 AHMADI, G., KOH, S. L., and GOLDSCHMIDT, V. W., A theory of non-simple microfluids, Recent advances in Engineering Science, Vol. **5**, Gordon and Breach, 9 (1970).
19 AHMADI, G., KOH, S. L. and GOLDSCHMIDT, V. W., Mechanics of a second order microfluid, Iranian J. Sci. Tech. **1**, 233 (1971).
20 AHMADI, G., Theory of second order micropolar fluids, Bull. de L'Acad. Polonaise des Sci. **25**, 15 (1977).
21 AHMADI, G., Heat Conducting microfluids, Indian J. Tech. **16**, 15 (1978).
22 ALLEN, S. J. and KLINE, K. A., Lubrication theory for micropolar fluids, J. Appl. Mech. Trans. ASME **38** 646 (1971).
23 BALARAM, M. and SASTRI, V. U. K., Micropolar lubrication, J. Appl. Mech. **39**, 834 (1972).
24 AHMADI, G., Lubrication theory for Cosserat fluids, (To appear).
25 AGRAWAL, V. K., GANJU, K. L. and JETHI, S. C., squeeze film and externally pressurized bearings micropolar lubricated, Wear **19**, 259 (1972).
26 PRAKASH, J. and SINNA, P., Squeeze film theory for micropolar fluids, J. Lubrication Technology Trans. ASME **97**, 139 (1976).
27 NEEDS, S. J., Boundary film investigations, Tran. ASME **62**, 331 (1940).

Eingereicht am 31. 5. 1980

Anschrift: Prof. GOODARZ AHMADI, Shiraz University, Shiraz, Iran, School of Engineering, Department of Mechanical Engineering and University of Calgary, Calgary, Canada, Department of Mechanical Engineering.

ZAMM 61, 221 – 227 (1981)

J. FÖRSTE

Existenz- und Einzigkeitsprobleme für instationäre Strömungen in offenen Gerinnen

Für sehr tiefe und sehr breite Rechteckgerinne wird die zeitliche Entwicklung einer Wasserwelle betrachtet. Es werden Existenz-, Einzigkeits- und Stabilitätsfragen untersucht.

The development of a water-wave is considered in a very deep or very wide open channel with rectangular cross-section. Problems of existence, uniqueness and stability are investigated.

Мы посмотрим развитие волн в одном открытом водном канале прямоугольного поперечного сечения. Исследуются проблемы существования, единственности и стабилизации решений.

Problemstellung

Wir betrachten die Wellenausbreitung in einem extrem tiefen und einem extrem breiten Rechteckgerinne der Länge $l = 1$. In einem sehr tiefen Gerinne hängt der hydraulische Radius nicht explizit vom durchflossenen Querschnitt ab; er kann gleich der halben Gerinnebreite b gesetzt werden, während er in einem sehr breiten Gerinne Querschnitt durch Breite $A/2b$ ist. Für laminare Strömung kann ein lineares Fließgesetz zugrunde gelegt werden, während für turbulente Strömungen die (quadratische) MANNING-STRICKLER-Formel zur Verfügung steht. Bezeichnet $u(x, t)$ die mittlere Fließgeschwindigkeit im Querschnitt an der Stelle x zur Zeit t, so gelten die von DE SAINT-VENANT aufgestellten Gleichungen

$$\frac{\partial A}{\partial t} + \frac{\partial Q}{\partial x} = q \,, \tag{1}$$

$$A\left(\frac{\partial u}{\partial t} + u\frac{\partial u}{\partial x}\right) + gA\frac{\partial h}{\partial x} = \nu\frac{\partial^2 u}{\partial x^2} + gA(I_s - I_e) - q(u - u_q\cos\alpha) \,, \tag{2}$$

wobei $Q = Au$ die pro Zeiteinheit durch den Querschnitt A fließende Wassermenge, $h = A/2b$ die Wassertiefe, g die Erdbeschleunigung, I_s das Sohlliniengefälle und ν die kinematische Zähigkeit des Wassers ist (vgl. [1], [2]). Für das Energieliniengefälle I_e hat man im laminaren Fall

$$I_e = \frac{3\nu}{gb^2}u \tag{3a}$$

für ein extrem tiefes Gerinne und

$$I_e = \frac{3\nu(2b)^2}{g}\cdot\frac{u}{A^2} \tag{3b}$$

für ein sehr breites Gerinne, während im turbulenten Fall die Formel

$$I_e = k_M^{-2}b^{-4/3}\,|u|\,u \tag{4a}$$

für ein sehr tiefes und

$$I_e = k_M^{-2}(2b)^{4/3}\,A^{-4/3}\,|u|\,u \tag{4b}$$

für ein extrem breites Gerinne verwendet wird. Dabei hängt der Beiwert k_M vom Material der Gerinnewand ab, q bedeutet den seitlichen Zufluß pro Längeneinheit und u_q die Geschwindigkeit des seitlich zufließenden Wassers, $\alpha(x)$ seinen Winkel mit der Flußrichtung im Gerinne.

Damit A und u durch (1) und (2) festgelegt werden, hat man Anfangs- und Randbedingungen zu stellen, wir verlangen

$$A(x, 0) = A_0(x)\,, \qquad u(x, 0) = u_0(x)\,, \qquad 0 \leqq x \leqq 1\,, \tag{5}$$

$$u(0, t) = u(1, t) = 0\,, \qquad t \geqq 0\,, \tag{6}$$

wobei u_0 und A_0 gegebene Funktionen sind und A_0 eine positive untere und obere Grenze m_0 und M_0 besitzt. Dies entspricht einem Gerinne, welches zwei auf unterschiedlichem Niveau befindliche Wasserreservoirs miteinander verbindet.

Gegenstand der vorliegenden Betrachtung sind Existenz- und Einzigkeitsaussagen für die Lösung der Anfangs-Randwertaufgabe (1), (2), (5), (6). Im laminaren und turbulenten Fall läßt sich die Einzigkeit der Lösung für ein beliebiges Zeitintervall $[0, T]$ zeigen; die Existenz können wir für ein Zeitintervall beweisen, dessen Länge von den gegebenen Größen abhängt. Im laminaren Fall läßt sich bei Abwesenheit von Nebenflüssen die Existenz auf ein beliebiges Zeitintervall $[0, T]$ ausdehnen. Verschwindet überdies das Sohlliniengefälle, so klingt jede Lösung mit der Zeit exponentiell ab (vgl. [3], [4], [5], [6]).

Einzigkeit der Lösung im turbulenten Fall

Angenommen, A', u' und A'', u'' seien zwei verschiedene Lösungen des Anfangs-Randwertproblems

$$\frac{\partial A}{\partial t} + \frac{\partial}{\partial x}(Au) = q\,, \tag{1}$$

$$A\left(\frac{\partial u}{\partial t} + u\frac{\partial u}{\partial x} + g\frac{\partial h}{\partial x}\right) = gA(I_s - k_M^{-2}b^{-4/3}|u|\,u) + \nu\frac{\partial^2 u}{\partial x^2} - q(u - u_q\cos\alpha); \tag{2a}$$

$$A(x, 0) = A_0(x)\,, \qquad u(x, 0) = u_0(x); \qquad u(0, t) = u(1, t) = 0\,, \tag{5), (6}$$

wobei $A', A'' \in W^{1,1}_{4,\infty}(Q_T)$, $A', A'' \geqq a > 0$; $u', u'' \in W^{2,1}_4(Q_T)$; $Q_T = [0, 1] \times [0, T]$ ist. Dabei beschreibt (2a) das tiefe Gerinne, für das breite Gerinne ist das erste Glied auf der rechten Seite durch $gA(I_s - k_M^{-2}(2b)^{4/3}A^{-4/3}|u|\,u)$ zu ersetzen.

Dann gilt für die Differenz der beiden Lösungen

$$A = A' - A''\,, \qquad u = u' - u'':$$

$$\frac{\partial A}{\partial t} + \frac{\partial}{\partial x}(Au' + A''u) = 0\,, \tag{7}$$

$$A'\left(\frac{\partial u}{\partial t} + u'\frac{\partial u}{\partial x} + u\frac{\partial u''}{\partial x}\right) + A\left(\frac{\partial u''}{\partial t} + u''\frac{\partial u''}{\partial x}\right) - \nu\frac{\partial^2 u}{\partial x^2} + \frac{g}{4b}\frac{\partial}{\partial x}(A'^2 - A''^2) - gI_sA +$$
$$+ gk_M^{-2}b^{-4/3}[Au'^2\operatorname{sgn}u' + A''(u' + u'')\,u\operatorname{sgn}u' + A''u''^2(\operatorname{sgn}u' - \operatorname{sgn}u'')] + qu = 0 \tag{8}$$

und A und u erfüllen homogene Anfangs- bzw. Randbedingungen. Für das sehr breite Gerinne ist in (8) das vorletzte Glied auf der linken Seite durch

$$k_M^{-2}(2b)^{4/3}(A'^{-1/3}\operatorname{sgn}u'\cdot u'^2 - A''^{-1/3}\operatorname{sgn}u''\cdot u''^2)$$
$$= k_M^{-2}(2b)^{4/3}(A'^{-1/3} - A''^{-1/3})\operatorname{sgn}u'\cdot u'^2 + A''^{-1/3}\operatorname{sgn}u'(u'^2 - u''^2) + A''^{-1/3}u''^2(\operatorname{sgn}u' - \operatorname{sgn}u'')]$$

zu ersetzen. Wir multiplizieren (7) mit A und (8) mit u und integrieren beide über x von 0 bis 1. Wenn wir

$$\left|\int_0^1 A\frac{\partial}{\partial x}(Au')\,dx\right| = \frac{1}{2}\left|\int_0^1 A^2\frac{\partial u'}{\partial x}\,dx\right| \leqq \frac{1}{2}\sup\left|\frac{\partial u'}{\partial x}\right|\int_0^1 A^2\,dx\,,$$

$$\left|\int_0^1 A\frac{\partial}{\partial x}(A''u)\,dx\right| \leqq \left|\int_0^1 Au\frac{\partial A''}{\partial x}\,dx\right| + \left|\int_0^1 AA''\frac{\partial u}{\partial x}\,dx\right| \leqq$$

$$\leqq 2\varepsilon\int_0^1\left(\frac{\partial u}{\partial x}\right)^2 dx + \frac{1}{\varepsilon^{3/5}}\left(\int_0^1\left(\frac{\partial A''}{\partial x}\right)^4 dx\right)^{2/5}\int_0^1(A^2 + u^2)\,dx + \frac{1}{\varepsilon}(\sup A'')^2\int_0^1 A^2\,dx\,,$$

$$\int_0^1 A'\left(\frac{\partial u}{\partial t} + u'\frac{\partial u}{\partial x}\right)u\,dx = \frac{1}{2}\frac{d}{dt}\int_0^1 A'u^2\,dx\,,$$

$$\left|\int_0^1\frac{g}{4b}\frac{\partial}{\partial x}(A'^2 - A''^2)\,u\,dx\right| = \left|\int_0^1\frac{g}{4b}(A'^2 - A''^2)\frac{\partial u}{\partial x}\,dx\right| \leqq$$

$$\leqq \frac{g}{4b}\sup(A' + A'')\left(\frac{\varepsilon}{2}\int_0^1\left(\frac{\partial u}{\partial x}\right)^2 dx + \frac{1}{2\varepsilon}\int_0^1 A^2\,dx\right),$$

$\varepsilon > 0$ beliebig, und

$$\left|\int_0^1 gk_M^{-2}b^{-4/3}[Au'^2\operatorname{sgn}u' + A''(u' + u'')\operatorname{sgn}u'\cdot u + A''u''^2(\operatorname{sgn}u' - \operatorname{sgn}u'')]\,u\,dx\right| \leqq$$
$$\leqq gk_M^{-2}b^{-4/3}\left[\tfrac{1}{2}\sup u'^2\int_0^1(A^2 + u^2)\,dx + \sup A''\sup|u' + u''|\int_0^1 u^2\,dx + 2\sup A''\sup u''^2\int_0^1 u^2\,dx\right]$$

bzw.

$$\left|\int_0^1 k_M^{-2}(2b)^{4/3}[(A'^{-1/3} - A''^{-1/3})\operatorname{sgn}u'\cdot u'^2 + A''^{-1/3}\operatorname{sgn}u'\cdot(u'^2 - u''^2) +\right.$$
$$\left. + A''^{-1/3}u''^2(\operatorname{sgn}u' - \operatorname{sgn}u'')]\,u\,dx\right| \leqq k_M^{-2}(2b)^{4/3}\left[\tfrac{1}{6}a^{-2/3}\sup u'^2\int_0^1(A^2 + u^2)\,dx +\right.$$
$$\left. + a^{-1/3}\sup|u' + u''|\int_0^1 u^2\,dx + 2a^{-1/3}\sup u''^2\int_0^1 u^2\,dx\right]$$

berücksichtigen, erhalten wir nach Addition beider Integralrelationen eine Ungleichung der Form

$$\int_0^1 (A^2 + u^2)\,dx \leqq \int_0^t f(t') \int_0^1 (A^2 + u^2)\,dx\,dt',$$

$0 \leqq t \leqq T$ mit integrierbarem $f(t)$, aus der mit Hilfe des GRONWALLschen Lemmas die Einzigkeit der Lösung folgt.

Ein Existenzsatz für den turbulenten Fall

Mit Hilfe eines Verfahrens der schrittweisen Näherung (vgl. [5], [3]) werden wir die Existenz einer Lösung des Anfangs-Randwertproblems (1), (2), (5), (6) für ein Zeitintervall $(0, t)$ beweisen, dessen Länge allerdings von den Ausgangsgrößen abhängen kann.

Zur Konstruktion von Näherungslösungen betrachten wir das lineare Problem

$$\frac{\partial A_k}{\partial t} + \frac{\partial(A_k u_{k-1})}{\partial x} = q, \qquad A_k|_{t=0} = A_0(x); \tag{9}$$

$$A_k \frac{\partial u_k}{\partial t} - \nu \frac{\partial^2 u_k}{\partial x^2} = -A_k g \frac{\partial h_k}{\partial x} - A_k u_{k-1} \frac{\partial u_k}{\partial x} + g A_k (I_s - u_k |u_{k-1}| k_M^{-2} b^{-4/3}) - q(u_k - u_q \cos\alpha) := f_k,$$

$$u_k|_{t=0} = u_0(x), \qquad u_k(0, t) = u_k(1, t) = 0 \tag{10a}$$

für das tiefe Gerinne, während für das breite Gerinne in (10a) $b^{-4/3}$ durch $(2b)^{4/3} A_k^{-4/3}$ zu ersetzen ist.

Wenn die nullte Näherung identisch 0 in $Q_T = (0, 1) \times (0, T)$ gesetzt wird, ist dieses Problem für beliebiges $k = 1, 2, \ldots$ und $T > 0$ in $W^{1,1}_{4,\infty}(Q_T)$ bzw. $W^{2,1}_4(Q_T)$ eindeutig lösbar. Als erstes suchen wir eine von k unabhängige Schranke für die Näherungslösengen. Für u_k gilt

$$U_k(t) := (||u_k||^{(2,1)}_{4,Q_t})^4 + \sup_{\tau \leqq t} (||u_k(x,\tau)||^{(3/2)}_4)^4 \leqq$$

$$\leqq C \frac{M_k^5(t)}{m_k^5(t)} \left(1 + m_k^{-1}(t) \sup_{\tau \leqq t} [A_k(x,\tau)]^{(\alpha)}\right)^{8/\alpha} \left(||f_k||^4_{4,Q_t} + (||u_0||^{(3/2)}_4)^4 + ||u_k||^4_{4,Q_t}\right) \tag{11}$$

mit $M_k(t) = \max\left(1, \max_{Q_t} A_k\right)$, $m_k(t) = \min\left(1, \min_{Q_t} A_k\right)$, wobei sich f_k wie folgt abschätzen läßt:

$$||f_k||^4_{4,Q_t} \leqq c M_k^4(t) \left(g^4 ||I_s||^4_{4,Q_t} + g^4 \int_0^t |u_{k-1}|^4 \left\| \frac{\partial u_k}{\partial x} \right\|_4^4 dt' + k_M^{-8} b^{-16/3} g^4 \int_0^t |u_{k-1}|^4 ||u_k||^4 dt' \right) +$$

$$+ \int_0^t q^4 ||u_k||_4^4 dt' + \int_0^t q^4 u_q^4 \cos^4\alpha\, dt' + \left(\frac{g}{2b}\right)^4 M_k^4(t) \int_0^t \left\| \frac{\partial A_k}{\partial x} \right\|_4^4 dt'; \tag{12a}$$

für das breite Gerinne ist darin $b^{-16/3}$ durch $(2b)^{16/3} [m_k(t)]^{-4/3}$ zu ersetzen.

Da mit den Charakteristiken $y(\tau, t, x)$ für $A_k(x, t)$ die Darstellung

$$A_k(x,t) = \exp\left(-\int_0^t \frac{\partial u_{k-1}}{\partial y} d\tau\right) \left\{ A_0(y(0,t,x)) + \int_0^t q \exp\left(\int_0^\tau \frac{\partial u_{k-1}}{\partial y} d\tau'\right) d\tau \right\}$$

gilt (vgl. [7]) hat man Ungleichungen der Form

$$m_0 \exp\left(-\int_0^t \left|\frac{\partial u_{k-1}}{\partial x}\right| d\tau\right) \leqq A_k(x,t) \leqq (M_0 + |q|\,T) \exp\left(\int_0^t \left|\frac{\partial u_{k-1}}{\partial x}\right| d\tau\right), \tag{13}$$

$$\left\|\frac{\partial A_k}{\partial x}\right\|_4 \leqq \exp\left(\int_0^t \left|\frac{\partial u_{k-1}}{\partial x}\right| d\tau\right) \left(\left\|\frac{\partial A_0}{\partial x}\right\|_4 + (M_0 + |q|\,T) \cdot \int_0^t \left\|\frac{\partial^2 u_{k-1}}{\partial x^2}\right\|_4 d\tau \right), \tag{14}$$

$$[A_k]^{(\alpha)} \leqq \left([A_0]^{(\alpha)} + (M_0 + |q|\,T) \int_0^t \left[\frac{\partial u_{k-1}}{\partial x}\right]^{(\alpha)} d\tau \right) \exp\left((1+\alpha) \int_0^t \left|\frac{\partial u_{k-1}}{\partial x}\right| d\tau \right). \tag{15}$$

Berücksichtigt man, daß sich alle auf den rechten Seiten von (11) bis (15) auftretenden Normen von u_k und u_{k-1} auf Grund der Einbettungssätze durch $U_k(t)$ bzw. $U_{k-1}(t)$ abschätzen lassen, so gewinnt man aus (11) eine Ungleichung der Form

$$U_k(t) \leqq G(R, t, U_{k-1}(t)) + H(R, t, U_{k-1}(t)) \int_0^t (U_{k-1}(t') + 1)\, U_k(t')\,dt',$$

wobei

$$R = |q|^4 + (||u_0||^{(3/2)}_4)^4 + \left\|\frac{\partial A_0}{\partial x}\right\|_4^4$$

gesetzt wurde. Aus dieser folgt

$$U_k(t) \leqq G \cdot \exp\left(H \cdot \left(t + \int_0^t U_{k-1}(t')\,\mathrm{d}t'\right)\right),$$

so daß man mit einer in allen drei Variablen monotonen Funktion K hat

$$U_k(t) \leqq K(R, t, U_{k-1}(t))\,.$$

Mittels vollständiger Induktion gewinnt man daraus die gewünschte a priori Abschätzung für die u_k. (13)—(15) liefern dann Abschätzungen der A_k. Wie in [5] oder [3] läßt sich damit beweisen, daß für $k \to \infty$ Grenzfunktionen vorhanden sind, die eine Lösung des Ausgangsproblems darstellen.

A priori Abschätzung und asymptotisches Verhalten für große t im laminaren Fall

Die Einzigkeits- und Existenzaussagen für die Anfangs-Randwertaufgabe (1), (2), (5), (6) gelten auch, wenn für das Energieliniengefälle die Formeln (3) zugrunde gelegt werden. Setzen wir zusätzlich voraus, daß keine Nebenflüsse vorhanden sind, so vereinfacht sich das System (1), (2) zu

$$\frac{\partial A}{\partial t} + \frac{\partial (Au)}{\partial x} = 0\,, \qquad A\left(\frac{\partial u}{\partial t} + u\frac{\partial u}{\partial x}\right) + \frac{\partial}{\partial x}\left(\frac{g}{4b}A^2\right) = \nu\frac{\partial^2 u}{\partial x^2} + gA\left(I_s - \frac{3\nu}{gb^2}u\right) \qquad (16), (17\,\mathrm{a})$$

im Fall des sehr tiefen Gerinnes und im Fall des sehr breiten Gerinnes ist $\frac{3\nu}{b^2}Au$ durch $\frac{3\nu(2b)^2}{A}u$ zu ersetzen.

Mit $v = 1/A$ wird aus (16), (17) bei gleichzeitigem Übergang zu LAGRANGEschen Koordinaten y, t

$$\frac{\partial u}{\partial y} - \frac{\partial v}{\partial t} = 0\,, \qquad \frac{\partial u}{\partial t} + \frac{\partial p}{\partial y} = \mu\frac{\partial}{\partial y}\left(\frac{1}{v}\frac{\partial u}{\partial y}\right) + g\left(I_s - \frac{3\mu}{gb^2}u\right), \qquad p = \frac{g}{4b}\frac{1}{v^2} \qquad (18), (19\,\mathrm{a})$$

bzw. $3\mu(2b)^2 v^2 u$ statt $\frac{3\mu}{b^2}u$. Dabei sind die Randbedingungen (6) für $y = 0$ und $y = d = \int_0^1 \varrho A_0(x)\,\mathrm{d}x$ zu stellen; ϱ bezeichnet die Dichte des Wassers und $\mu = \varrho\nu$ seine dynamische Zähigkeit. Die Anfangsbedingungen (5) bleiben unverändert. Wenn wir

$$\varphi^2(v) = \frac{g}{4b}\int_1^v \left(1 - \frac{1}{s^2}\right)\mathrm{d}s > 0$$

setzen und (19) in $L^2(0, d)$ skalar mit u multiplizieren, erhalten wir

$$\frac{\mathrm{d}}{\mathrm{d}t}\int_0^d \left[\frac{1}{2}u^2 + \varphi^2(v)\right]\mathrm{d}y + \mu\int_0^d \frac{1}{v}\left(\frac{\partial u}{\partial y}\right)^2\mathrm{d}y + \frac{3\mu}{b^2}\int_0^d u^2\,\mathrm{d}y = \int_0^d gI_s u\,\mathrm{d}y \qquad (20)$$

bzw. an Stelle des letzten Gliedes auf der linken Seite in (20) $3\mu(2b)^2\int_0^d v^2u^2\,\mathrm{d}y$ für das sehr breite Gerinne. Da sich (19) unter Verwendung von (18) in der Form

$$\mu\frac{\partial}{\partial t}\frac{\partial \ln v}{\partial y} = \frac{\partial u}{\partial t} + \frac{\partial p}{\partial y} + \frac{3\mu}{b^2}u - gI_s \qquad (21)$$

bzw. $3\mu(2b)^2 v^2 u$ statt $\frac{3\mu}{b^2}u$ schreiben läßt, gewinnt man durch skalare Multiplikation mit $\frac{\partial \ln v}{\partial y}$ in $L^2(0, d)$ aus (21)

$$\frac{\mu}{2}\frac{\mathrm{d}}{\mathrm{d}t}\int_0^d \left(\frac{\partial \ln v}{\partial y}\right)^2\mathrm{d}y = \int_0^d \frac{\partial u}{\partial t}\frac{\partial \ln v}{\partial y}\mathrm{d}y + \int_0^d vp'(v)\left(\frac{\partial \ln v}{\partial y}\right)^2\mathrm{d}y + \frac{3\mu}{b^2}\int_0^d u\frac{\partial \ln v}{\partial y}\mathrm{d}y - \int_0^d gI_s\frac{\partial \ln v}{\partial y}\mathrm{d}y \quad (22)$$

bzw. $3\mu(2b)^2\int_0^d uv^2\frac{\partial \ln v}{\partial y}\,\mathrm{d}y$ statt des vorletzten Integrals in (22).

Wegen

$$\frac{\mathrm{d}}{\mathrm{d}t}\int_0^d u\frac{\partial \ln v}{\partial y}\mathrm{d}y = \int_0^d \frac{\partial u}{\partial t}\frac{\partial \ln v}{\partial y}\mathrm{d}y - \int_0^d \frac{1}{v}\left(\frac{\partial u}{\partial y}\right)^2\mathrm{d}y\,,$$

$$\int_0^d u\frac{\partial \ln v}{\partial y}\mathrm{d}y = -\frac{\mathrm{d}}{\mathrm{d}t}\|\psi(v)\|^2\,, \qquad \psi^2(v) = v\ln v - v + 1$$

und $p'(v) = -\frac{g}{2b}\frac{1}{v^3}$ kann (22) auch in der Form

$$\frac{\mathrm{d}}{\mathrm{d}t}\left\{\frac{\mu}{2}\int_0^d\left(\frac{\partial \ln v}{\partial y}\right)^2 \mathrm{d}y + \frac{3\mu}{b^2}\,||\psi(v)||^2\right\} + \frac{g}{2b}\int_0^d \frac{1}{v^2}\left(\frac{\partial \ln v}{\partial y}\right)^2 \mathrm{d}y$$
$$= \frac{\mathrm{d}}{\mathrm{d}t}\int_0^d u\,\frac{\partial \ln v}{\partial y}\,\mathrm{d}y + \int_0^d \frac{1}{v}\left(\frac{\partial u}{\partial y}\right)^2 \mathrm{d}y - \int_0^d g I_s\,\frac{\partial \ln v}{\partial y}\,\mathrm{d}y \tag{23}$$

bzw. $\frac{1}{2}\,\mu(2b)^2 \int_0^d v^3\,\mathrm{d}y$ statt $\frac{3\mu}{b^2}\,||\psi(v)||^2$ geschrieben werden. Nach Integration über t von 0 bis T folgt aus (23) unter Verwendung von (20), der Ungleichung $ab \leqq \frac{\varepsilon}{2}\,a^2 + \frac{1}{2\varepsilon}\,b^2$ und des GRONWALLschen Lemmas

$$\int_0^d \left(\frac{\partial \ln v}{\partial y}\right)^2 \mathrm{d}y \leqq M(T)\,.$$

Ist $I_s = 0$, so hängt die Konstante M nicht von der Zeit ab.

Nach den Einbettungssätzen ist dann $\ln v$ auf $(0, d)$ gleichmäßig beschränkt, so daß für v positive untere und obere Schranken existieren $0 < N^{-1} \leqq v \leqq N$.

Nach (20) ist dann auch u gleichmäßig beschränkt. Entsprechendes läßt sich für die Ableitungen beweisen (vgl. [6]). Im weiteren setzen wir zusätzlich voraus, daß das Sohlliniengefälle I_s verschwindet. Um das Verhalten von u und v für große t zu untersuchen, multiplizieren wir (20) mit $\gamma > \mu^{-1}$ und addieren danach (20) zu (23). Wir haben dann

$$\frac{\mathrm{d}^2}{\mathrm{d}t^2}\,||\psi(v)||^2 + \frac{\mathrm{d}}{\mathrm{d}t}\,U(t) + V(t) = 0 \tag{24}$$

mit

$$U(t) = \frac{\gamma}{2}\,||u||^2 + \gamma\,||\varphi(v)||^2 + \frac{\mu}{2}\left\|\frac{\partial \ln v}{\partial y}\right\|^2 + \frac{3\mu}{b^2}\,||\psi(v)||^2$$

bzw. $$\frac{1}{2}\,\mu(2b)^2 \int_0^d v^3\,\mathrm{d}y \text{ statt } \frac{3\mu}{b^2}\,||\psi(v)||^2;$$

$$V(t) = (\mu\gamma - 1)\int_0^d \frac{1}{v}\left(\frac{\partial u}{\partial y}\right)^2 \mathrm{d}y + \frac{g}{2b}\int_0^d \frac{1}{v^2}\left(\frac{\partial \ln v}{\partial y}\right)^2 \mathrm{d}y + \frac{3\mu\gamma}{b^2}\int_0^d u^2\,\mathrm{d}y$$

bzw. $3\mu\gamma(2b)^2 \int_0^d v^2u^2\,\mathrm{d}y$ statt des letzten Summanden.

Wegen

$$\frac{\gamma}{2}\,||u||^2 \leqq \frac{\gamma N}{2}\int_0^d \frac{1}{v}\left(\frac{\partial u}{\partial y}\right)^2 \mathrm{d}y\,, \qquad \gamma||\varphi(v)||^2 \leqq \frac{\gamma g N}{4b}\int_0^d \frac{1}{v^2}\left(\frac{\partial \ln v}{\partial y}\right)^2 \mathrm{d}y\,,$$

$$\frac{\mu}{2}\left\|\frac{\partial \ln v}{\partial y}\right\|^2 \leqq \frac{\mu N^2}{2}\int_0^d \frac{1}{v^2}\left(\frac{\partial \ln v}{\partial y}\right)^2 \mathrm{d}y\,, \qquad \psi^2(v) \leqq N^2\varphi^2(v)$$

und

$$\frac{3\mu}{b^2}\,||\psi(v)||^2 \leqq \frac{3\,\mu g N^3}{4b^3}\int_0^d \frac{1}{v^2}\left(\frac{\partial \ln v}{\partial y}\right)^2 \mathrm{d}y$$

bzw.

$$\frac{1}{2}\,\mu(2b)^2 \int_0^d v^3\,\mathrm{d}y \leqq \frac{1}{2}\,\mu(2b)^2\,N^5 \int_0^d \frac{1}{v^2}\left(\frac{\partial \ln v}{\partial y}\right)^2 \mathrm{d}y$$

hat man

$$U(t) \leqq \frac{\gamma N}{2}\int_0^d \frac{1}{v}\left(\frac{\partial u}{\partial y}\right)^2 \mathrm{d}y + \left(\frac{\gamma g N}{4b} + \frac{\mu N^2}{2} + \frac{3\mu g N^3}{4b^3}\right)\int_0^d \frac{1}{v^2}\left(\frac{\partial \ln v}{\partial y}\right)^2 \mathrm{d}y$$
$$= \frac{N}{2}\,\frac{2b}{g}\left(\frac{\gamma g}{2b} + \mu N + \frac{3\mu g N^2}{2b^3}\right)\left[\frac{\frac{\gamma g}{2b}}{\frac{\gamma g}{2b} + \mu N + \frac{3\mu g N^2}{2b^3}}\int_0^d \frac{1}{v}\left(\frac{\partial u}{\partial y}\right)^2 \mathrm{d}y + \frac{g}{2b}\int_0^d \frac{1}{v^2}\left(\frac{\partial \ln y}{\partial y}\right)^2 \mathrm{d}y\right],$$

wobei im Falle des sehr breiten Gerinnes hier und im folgenden $\frac{3\mu g N^2}{2b^3}$ durch $\mu(2b)^2 N^4$ zu ersetzen ist.

Für $\mu\gamma - 1 = \frac{\gamma \frac{g}{2b}}{\frac{\gamma g}{2b} + \mu N + \frac{3\mu g N^2}{2b^3}}$ erhält man

$$U(t) \leqq \frac{N}{2} \cdot \frac{2b}{g} \left(\frac{\gamma g}{2b} + \mu N + \frac{3\mu g N^2}{2b^3}\right) \left(V(t) - \frac{3\mu\gamma}{b^2} \int_0^d u^2 \, dy \right),$$

folglich

$$V(t) \geqq \delta U(t) \quad \text{mit} \quad \delta = \frac{g}{bN} \frac{1}{\frac{\gamma g}{2b} + \mu N + \frac{3\mu g N^2}{3b^3}}.$$

Zur Bestimmung von γ entsteht die quadratische Gleichung

$$\mu g b^2 \gamma^2 + (2b^3\mu^2 N + 3\mu^2 g N^2 - 2gb^2)\gamma - 2b^2\mu N - 3\mu g N^2 = 0\,,$$

aus der sich

$$\gamma = \frac{-(2b^3\mu^2 N + 3\mu^2 g N^2 - 2gb^2) + \sqrt{(2b^3\mu^2 N + 3\mu^2 g N^2 - 2gb^2)^2 + 4\mu g b^2 (2b^2\mu N + 3\mu g N^2)}}{2\mu g b^2}$$

ergibt, so daß γ positiv ist und groß wird, wenn N groß ist. Für ein sehr breites Gerinne hat man

$$\mu\gamma - 1 = \frac{\frac{\gamma g}{2b}}{\frac{\gamma g}{2b} + \mu N + \mu (2b)^2 N^4},$$

$$\frac{\mu g}{2b}\gamma^2 + \left(\mu^2 N + \mu^2 (2b)^2 N^4 - \frac{g}{b}\right)\gamma - (\mu N + \mu(2b)^2 N^4) = 0\,,$$

also

$$\gamma = \frac{-(b^2\mu^2 N + 4\mu^2 b^3 N^4 - g) + \sqrt{(b^2\mu^2 N + 4\mu^2 b^3 N^4 - g)^2 + 2\mu g(\mu b N + 4\mu b^3 N^4)}}{\mu g}.$$

Aus (24) kann jetzt auf

$$e^{\delta t} \frac{d^2}{dt^2} ||\psi(v)||^2 + \frac{d}{dt} (e^{\delta t}\, U(t)) \leqq 0 \tag{25}$$

geschlossen werden, woraus

$$U(t)\, e^{\delta t} + \int_0^t e^{\delta\tau} \frac{d^2}{d\tau^2} ||\psi(v)||^2 \, d\tau \leqq U(0) \tag{26}$$

folgt. Da

$$\int_0^t e^{\delta\tau} \frac{d^2}{d\tau^2} ||\psi(v)||^2 \, d\tau = e^{\delta t} \frac{d}{dt} ||\psi(v)||^2 - \frac{d}{dt} ||\psi(v)||^2 \Big|_{t=0} - \delta\, e^{\delta t}\, ||\psi(v)||^2 + \delta ||\psi(v)||^2 \Big|_{t=0} + \delta^2 \int_0^t e^{\delta\tau}\, ||\psi(v)||^2 \, d\tau$$

gilt, läßt sich (26) in der Form

$$U(t) - \int_0^d u \frac{\partial \ln v}{\partial y} \, dy - \delta ||\psi(v)||^2 \leqq C\, e^{-\delta t}$$

schreiben, wo C nur von gegebenen Größen abhängt. Verwendet man

$$\left| \int_0^d u \frac{\partial \ln v}{\partial y} \, dy \right| \leqq \frac{\varepsilon}{2} \int_0^d \left(\frac{\partial \ln v}{\partial y}\right)^2 dy + \frac{1}{2\varepsilon} \int_0^d u^2 \, dy\,, \qquad \varepsilon > 0 \quad \text{beliebig}\,,$$

so erhält man für den Fall des sehr tiefen Gerinnes

$$\frac{3\mu}{b^2} ||\psi(v)||^2 + \frac{\gamma - \varepsilon^{-1}}{2} ||u||^2 + \gamma ||\varphi(v)||^2 + \frac{\mu - \varepsilon}{2} \left\| \frac{\partial \ln v}{\partial y} \right\|^2 - \delta ||\psi(v)||^2 \leqq C\, e^{-\delta t}\,, \tag{27a}$$

wobei durch hinreichend großes N erreicht werden kann, daß $\frac{3\mu}{b^2} - \delta > 0$ ist. Wegen $||\varphi(v)||^2 \leqq \frac{gN^3}{4b} \left\| \frac{\partial \ln v}{\partial y} \right\|^2$ erhalten wir

$$\tfrac{1}{2} ||u||^2 + ||\varphi(v)||^2 \leqq C\, e^{-\delta t}\,, \tag{28}$$

falls wir ε so wählen, daß

$$\gamma - \varepsilon^{-1} = \gamma + \frac{\mu - \varepsilon}{2} \frac{4b}{gN^3}, \quad \text{d. h.} \quad (\varepsilon^{-1})^2 + \frac{2b\mu}{gN^3}\varepsilon^{-1} - \frac{2b}{gN^3} = 0 ,$$

also

$$\varepsilon^{-1} = -\frac{b\mu}{gN^3} + \sqrt{\left(\frac{b\mu}{gN^3}\right)^2 + \frac{2b}{gN^3}}$$

ist. Das bedeutet, u und $v - 1$ klingen mit der Zeit exponentiell ab.

Bei einem extrem breiten Gerinne hat man an Stelle von (27a)

$$\frac{\mu b^2}{2} \int_0^d v^3 \,\mathrm{d}y + \frac{\gamma - \varepsilon^{-1}}{2} ||u||^2 + \gamma||\varphi(v)||^2 + \frac{\mu - \varepsilon}{2} \left\|\frac{\partial \ln v}{\partial y}\right\|^2 - \delta||\psi(v)||^2 \leqq C\,\mathrm{e}^{-\delta t} . \tag{27b}$$

Läßt man in (27b) das erste Integral weg, berücksichtigt

$$\psi^2(v) \leqq N^2\varphi^2(v) , \qquad ||\varphi(v)||^2 \leqq \frac{gN^3}{4b} \left\|\frac{\partial \ln v}{\partial y}\right\|^2$$

und verlangt

$$\gamma - \varepsilon^{-1} = \gamma + \frac{\mu - \varepsilon}{2} \frac{4b}{gN^3} - \delta N^2 ,$$

so kommt man ebenfalls auf (28). Man bestimmt ε^{-1} aus der quadratischen Gleichung

$$(\varepsilon^{-1})^2 + \left(\frac{2b\mu}{gN^3} - \delta N^2\right)\varepsilon^{-1} - \frac{2b}{gN^3} = 0$$

zu

$$\varepsilon^{-1} = \frac{1}{2}\left(\delta N^2 - \frac{2b\mu}{gN^3}\right) + \frac{1}{2}\sqrt{\left(\frac{2b\mu}{gN^3} - \delta N^2\right)^2 + \frac{8b}{gN^3}} .$$

Literatur

1 J. Edenhofer und Schmitz, G., Ein zeitschrittunabhängiges Charakteristikenverfahren zur Berechnung instationärer Strömungsvorgänge in offenen Gerinnen, ZAMM **60**, T 178—T 180 (1980).

2 Press, H. und Schröder, R., Hydromechanik im Wasserbau, Berlin und München 1966.

3 Förste, J., Über die Lösbarkeit der Feldgleichungen des Zweiflüssigkeitsmodells für ein staubbeladenes Gas, ZAMM **58**, 189—198 (1978).

4 А. В. Кажихов, А. Н. Петров, Корректность начально-краевой задачи для модельной системы уравнений многокомпонентной смеси, Динамика сплошной среды **35**, 61—73, СОАН СССР, Институт гидродинамики 1978г.

5 В. А. Солонников, О разрешимости начально-краевой задачи для уравнений движения вязкой сжимаемой жидкости, Записки научных семинаров ЛОМИ **56**, 128—142 (1976)

6 А. В. Кажихов, О стабилизации решений начально-краевой задачи для уравнений баротропной вязкой жидкости, Дифференциальные уравнения XV, № 4, 662—667 (1979)

7 Kamke, E., Differentialgleichungen Lösungsmethoden und Lösungen II, Leipzig 1956.

Eingereicht am 10. 11. 1980

Anschrift: Prof. Dr. Joachim Förste, Inst. für Mechanik der AdW, DDR-1199 Berlin, Rudower Chausee

ZAMM 61, 229–234 (1981)

K.-F. FISCHER

On an Analytical Determination of Initial Points of Crack Propagation

In der vorliegenden Arbeit wird eine Möglichkeit der Berechnung von Initialpunkten der Rißausbreitung von der Rißspitze untersucht. Diese besteht darin, bei der Formulierung charakteristischer Größen zur Rißbeurteilung neben dem ersten Glied der Reihenlösung für die elastischen Feldgrößen in Rißspitzenumgebung weitere Glieder zu berücksichtigen. Es wird vor allem ein derartig erweitertes Sih-Kriterium vorgestellt. Eine qualitativ neue Aussage bildet die Berechnung von hypothetischen Initialpunkten der Rißausbreitung vor der Rißspitze. Ähnliche Ergebnisse erhält man bei der Untersuchung eines erweiterten Spannungs-Parameter-Kriteriums und eines erweiterten Kriteriums der Mehrachsigkeitszahl des Spannungszustandes.

In the present paper, a possibility of determination of initial points of crack propagation in front of the crack tip is investigated. This possibility consists in the fact that apart from the first term of the series solution for the elastic field variables in crack tip vicinity further terms are considered in the formulation of characteristic quantities for crack judgment. Above all an extended Sih-Criterion is presented. A qualitatively new information results in the determination of hypothetical initial points of crack propagation in front of the crack tip. We obtain similar results if we consider the extended stress-parameter-criterion and the extended multi-axial-value-criterion.

В данной работе изучается возможность определения начальных точек распространения трещины перед остриём трещины. Эта возможность состоит в том, что, исключая первый член ряда решения для упругих переменных поля, в непосредственной близости от острия трещины дальнейшие члены рассматриваются с помощью построения характеристических величин для оценки трещин. Прежде всего, представлен расширенный критерий Си. Качественно новая информация представлена определением гипотетических начальных точек распространения трещины перед остриём трещины. Подобные результаты получаются, если рассмотреть расширенный критерий напряжения и расширенный критерий многоосевой величины.

1. Introduction

In literature, various fracture criteria of crack fracture mechanics are described, [1], [2], [3]. They are classified, for instance, according to the main physical quantities such as strain energy, stress, deformation. The form of this physical magnitude, that is specific for the particular criterion, is called a "characteristic quantity". Fracture criteria include information on the following two basic hypotheses [2], [4]:

1st hypothesis: Assumption about crack propagation direction in case of crack instability.

2nd hypothesis: The crack becomes unstable, if the characteristic quantity for crack judgment reaches a critical value.

In the following, we restrict ourselves to linearly elastic material behaviour and static, in-plane problems.

The stress and displacement field in the crack tip vicinity provides the basis for the determination of the characteristic quantities. The stress field takes the following form approximately:

$$\overset{k}{\sigma}_{ij} = \sum_{n=1}^{k} r^{n/2-1} \cdot f_{ij}(a_n^R, a_n^I, \vartheta) . \tag{1}$$

μ and ν are the shear modulus and POISSON's ratio respectively. The a_n^R, a_n^I denote the generalized intensity factors. With $n = 1$, the $r^{-1/2}$-singularity characteristic for crack problems is obtained.

In most cases only the terms for $n = 1$ of the stress field in the crack tip vicinity are considered for formulating criteria of crack fracture mechanics at present. For improved clarify such criteria will be refered to as "simple" criteria in the following.

Now a survey on the formal extension of fracture criteria will be given in such a way that in addition to the first term further terms are considered in the formulation of characteristic quantities. Such criteria will be refered to as "extended" criteria of crack fracture mechanics, cf. [5].

COTTERELL was the first who underlined the significance of including further terms in formulating characteristic quantities, [6]. WILLIAMS and EWING, for instance, used an extended normal stress criterion with two terms to treat a plane GRIFFITH-crack under tension loading, [7].

In the following, the extended SIH-Criterion, the extended stress-parameter-criterion and the extended multiaxial-value-criterion will be discussed as examples for extended criteria and their significance for the determination of hypothetical initial points of crack propagation.

2. Extension of the Sih-Criterion

The strain energy density factor S is used as a characteristic quantity, [1], [8]. In general, S can be expressed by

$$S = rW^* . \tag{2}$$

W^* describes the strain energy density. In the following, ϑ and r describe the coordinates in a polar coordinate system with the origin in the crack tip.

The criterion is based upon the fact that the fracture proceeds from an failed "interior element" located in the vicinity of the crack tip. It is expedient to use a representation of W^* by means of the principal normal stresses:

$$W^* = \frac{1}{16\mu}\left[2(\sigma_1 - \sigma_2)^2 + (\varkappa - 1)(\sigma_1 + \sigma_2)^2\right], \tag{3}$$

where

$$\varkappa = \begin{cases} (3 - 4\nu), & \text{plane strain}, \\ (3 - \nu)/(1 + \nu), & \text{plane stress}. \end{cases}$$

The following sum and difference of principal normal stresses will result:

$$(\sigma_1 + \sigma_2)^2 = \sum_{n=1}^{k} b_n r^{(n-3)/2}, \qquad (\sigma_1 - \sigma_2)^2 = \sum_{n=1}^{k} c_n r^{(n-3)/2}. \tag{4}$$

The coefficients b_n, c_n depend upon ϑ and the generalized intensity factors.

Now we write the strain energy density factor (or function, respectively) $\overset{k}{S}$:

$$\overset{k}{S} = r\overset{k}{W}{}^*. \tag{5}$$

The superscript index indicates the number of the terms considered in determining the corresponding quantity. Using equations (3), (4) we obtain the decomposition:

$$\overset{k}{S} = \sum_{n=1}^{k} S_n, \qquad S_n = \frac{a^{(n-1)/2}}{16\mu}\left(\frac{r}{a}\right)^{(n-1)/2}\left[2c_n + (\varkappa - 1)\,b_n\right]. \tag{6}$$

For $r/a \to 0$ and $\lim\limits_{r/a \to 0} (r/a)^0 \doteq 1$, equation (6) changes into the wellknown $\overset{1}{S}$:

$$\overset{1}{S} = \frac{K_I^2}{16\mu\pi}\left(a_{11} + 2a_{12}M + a_{22}M^2\right).$$

The coefficients a_{ij} depend upon ϑ and POISSON's ratio ν. With $r/a \doteq \xi$, the following dependence results:

$$S_n = S_n(\vartheta, \xi). \tag{7}$$

According to the 1st hypothesis of the simple criterion, the locus of the local minimum of the strain energy density factor governs the crack propagation direction. Due to the dependence given in equation (7), the following relations are available for the determination of the point $\overset{k}{P}_0\left(\overset{k}{\vartheta}_0, \overset{k}{\xi}_0\right)$ in which $\overset{k}{S} = \overset{k}{S}_{\min}$:

$$\left\{\partial\overset{k}{S}/\partial\vartheta = 0, \qquad \partial\overset{k}{S}/\partial\xi = 0\right\}\Bigg|_{\substack{\vartheta = \overset{k}{\vartheta}_0 \\ \xi = \overset{k}{\xi}_0}}, \tag{8}$$

$$\left\{\left[\frac{\partial^2\overset{k}{S}}{\partial\vartheta^2}\cdot\frac{\partial^2\overset{k}{S}}{\partial\xi^2} - \left(\frac{\partial^2\overset{k}{S}}{\partial\vartheta\cdot\partial\xi}\right)^2\right] > 0, \qquad \frac{\partial^2\overset{k}{S}}{\partial\xi^2} > 0\right\}\Bigg|_{\substack{\vartheta = \overset{k}{\vartheta}_0 \\ \xi = \overset{k}{\xi}_0}}. \tag{9}$$

If there exists such a point, an extended SIH-Criterion can be stated:

1st hypothesis: Crack propagation in the elastic range proceeds from the point $\overset{k}{P}_0\left(\overset{k}{\vartheta}_0, \overset{k}{\xi}_0\right)$ radially towards $\overset{k}{\vartheta}_0$. At this point the strain energy density factor $\overset{k}{S}$ becomes a local minimum.

2nd hypothesis: The crack becomes unstable when the minimum $\overset{k}{S}$-factor reaches a critical value S_C.

$\overset{k}{P}_0$ is refered to as the Hypothetical Initial Point of the Crack Propagation, [9], [10], (see Fig. 1).

Now at first we base our considerations on the GRIFFITH-crack under Mode I-loading, $\overset{k}{\vartheta}_0 = 0^0$.

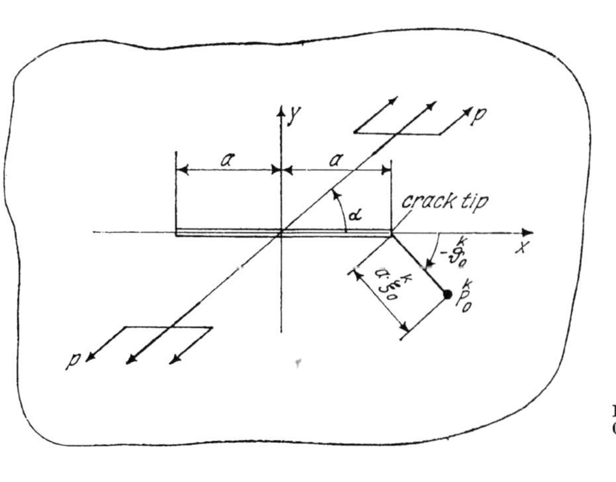

Fig. 1. Positions of Initial Points of Crack Propagation near the Crack Tip

If we take $k = 4$ into account from equations (8) result:

$$\overset{4}{\xi}_0 = \left\{\frac{2}{9(\varkappa - 1)}\left[\sqrt{2}\,(2\varkappa - 1) - \sqrt{\varkappa(10 - \varkappa) - 7}\right]\right\}^2. \tag{10}$$

For comparison one obtains:

$$\overset{3}{\xi}_0 = \frac{1}{2}\left(\frac{\varkappa - 1}{2\varkappa - 1}\right)^2, \tag{11}$$

$$\overset{\infty}{\xi}{}_0^{1/2}\left(\overset{\infty}{\xi}_0 + 2\right)^{5/2}[2 + 5(\varkappa - 1)] - (\varkappa - 1)\left\{4\overset{\infty}{\xi}{}_0^{1/2}\left(\overset{\infty}{\xi}_0 + 2\right)^{1/2} + 2\left(\overset{\infty}{\xi}_0 + 2\right) + \left[\left(\overset{\infty}{\xi}_0 + 2\right)\left(3\overset{\infty}{\xi}_0 + 1\right) - \overset{\infty}{\xi}_0\left(\overset{\infty}{\xi}_0 + 1\right)\right]\right\} = 0, \tag{12}$$

where

$$\begin{Bmatrix} \overset{\infty}{\sigma}_1 \\ \overset{\infty}{\sigma}_2 \end{Bmatrix} = p \begin{Bmatrix} (\xi + 1)/(\xi^2 + 2\xi)^{1/2} \\ (\xi + 1)/(\xi^2 + 2\xi)^{1/2} - 1 \end{Bmatrix}. \tag{13}$$

Figure 2 shows the dependence of the distance between crack tip and initial point upon Poisson's ratio.

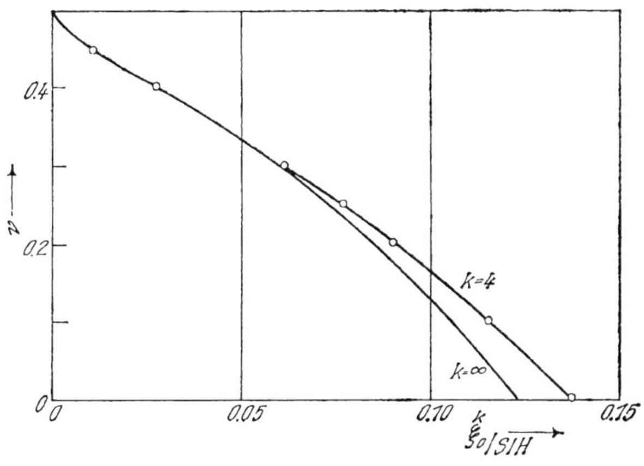

Fig. 2. Dependence of the distance between initial points and crack tip upon Poisson's ratio ν, Mode I

Now we get the following values for the critical load ($\nu = 0.3$):

$$\left(\overset{k}{p}_C\right)^2 \frac{a}{\mu \cdot S_C} = \begin{Bmatrix} 10, & k = 1 \\ 12.005, & k = 4 \\ 11.989, & k = \infty \end{Bmatrix}. \tag{14}$$

Thus the simple criterion can be understood as the lower limit for the critical load of the crack. The value $\overset{\infty}{p}_C$, which was determined according to the exact solution of the stress field at the crack tip (cf. equations (13)), shows that, with $k = 4$, a good approximation has been obtained.

Now again, we base our considerations on the angled Griffith-crack. Thus, the relations (8) lead to a non linear set of equations for the determination of the coordinates of the initial point:

$$\left\{\begin{aligned} &\sum_{n=1}^{k} a^{(n-1)/2}\overset{k}{\xi}{}_0^{(n-1)/2}\left[2\,\frac{\partial c_n}{\partial \vartheta} + (\varkappa - 1)\,\frac{\partial b_n}{\partial \vartheta}\right] = 0 \\ &\sum_{n=1}^{k} a^{(n-1)/2}\overset{k}{\xi}{}_0^{(n-3)/2}\left(\frac{n-1}{2}\right)[2c_n + (\varkappa - 1)\,b_n] = 0 \end{aligned}\right\}_{\vartheta = \overset{k}{\vartheta}_0}. \tag{15}$$

This set of coupled equations is solved numerically for $k = 4$ and $\nu = 0.3$ (plane strain) for various load application angles.

The extended criterion, in contrast to the simple criterion, leads — both for the crack propagation angle and the minimum energy density factor — to quantitatively differing results. Fig. 3 shows, for instance, the dependence of the crack propagation angle upon the load application angle.

It is necessary to mention that nearly the same results are obtained if we use the exact solution for the stress field in crack tip vicinity.

The qualitative difference between the simple and extended Sih-Criterion is expressed by the prediction of the existence of an initial point of crack propagation in front of the crack tip. The significance of the prediction of initial points will be discussed in brief.

First it should be said that the determination of such points is based solely on the formulation of an extended strain energy density factor (or, better: -function) and the minimization of the parameter. Test results are known from literature, where crack nucleus formation (or now: initial points) in front of the crack tip was observed in polycarbonate samples, [11]. In order to enable a comparison with the results obtained by means of analytic cal-

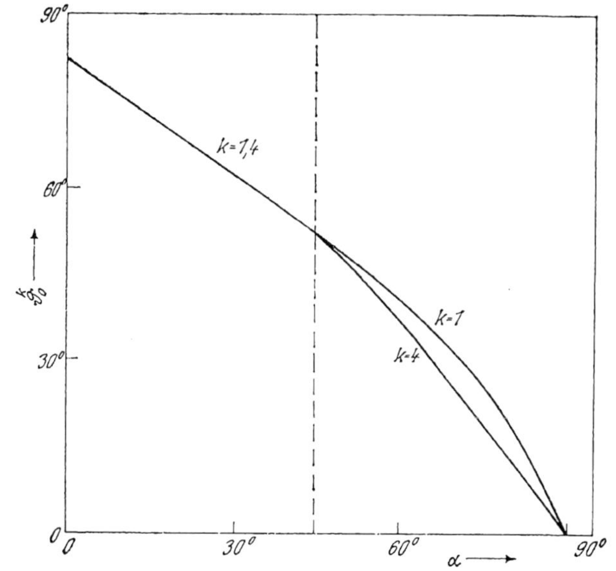

Fig. 3. Dependence of the crack propagation angle $\overset{k}{\vartheta}_0$ upon the load application angle α, $\nu = 0.3$ (plane strain)

Table 1. Positions of initial points according the extended SIH-Criterion

crack sharpness $\delta = \varrho/a$	ξ_0 according to GARDE/WEISS [11]	$\overset{\infty}{\xi}_0$ (solution for ellipse)	
		$\nu = 0.3$	$\nu = 0.4$
0.0032	0.00897	0.0718	0.0295
0.016	0.0311	0.0868	0.0349
0.032	0.0577	0.0972	0.0385

culations, it is necessary to consider the influence of the V-notch radius in formulating S. From this the values given in Table 1 result for the distance of the initial points from the roots of the notches for various "crack sharpness degress" $\delta = \varrho/a$ where ϱ denotes the notch radius.

Because of the difference between of the crack models (tension bar with double outer notch [11] and here ellipse in the infinite plane) only a qualitative comparison can be made. The theoretical prediction stating that ξ_0 increases with growing δ, is substantiated.

3. Extension of the Stress-Parameter-Criterion

WANG TZU CHIANG proposed in 1975 a so called stress parameter criterion for combined mode fracture, [12]. The simple criterion includes the following basic hypotheses:

1st hypothesis: Crack propagation proceeds in radial direction vertically to the maximum normal stress on an iso-W^*-line, i.e.

$$\left\{\partial \overset{1}{\Phi}/\partial\vartheta = 0,\ \partial^2\overset{1}{\Phi}/\partial\vartheta^2 < 0\right\}_{|\vartheta = \overset{1}{\vartheta}_0} \tag{16}$$

where $\overset{1}{\Phi}$ is the stress parameter

$$\overset{1}{\Phi} = \frac{K_I}{\sqrt{\pi \overset{1}{S}}}\left[1 + \cos\vartheta - 3M\sin\vartheta\right]\cos\frac{\vartheta}{2}\,. \tag{17}$$

2nd hypothesis: The crack becomes unstable when the maximum normal stress on an iso-W^*-line reaches a critical value:

$$\overset{1}{\sigma}_{\vartheta|\vartheta = \overset{1}{\vartheta}_0} \doteq \max \overset{1}{\sigma}_{\vartheta|\overset{1}{W}{}^* = \text{const.}} = \sigma_C\,. \tag{18}$$

Magnitude M is defined by $M \doteq K_{II}/K_I$. K_I and K_{II} are the stress intensity factors for symmetrical and antimetrical crack loading. Since magnitude M can be understood as a measure of deviation from the symmetrical case it is refered to as the symmetry factor, cf. [2], [5].

If we extend the criterion in the sense described above, and use the exact stress field around the crack tip of a GRIFFITH-crack under Mode I-loading, the stress parameter reads as follows, $\overset{k}{\vartheta}_0 = 0^0$:

$$\overset{\infty}{\Phi} = (1 + \xi)/\left\{(\xi + 2)\left[2\xi + (\varkappa - 1)\,\xi\left(\frac{2(1 + \xi)}{\sqrt{\xi^2 + 2\xi}} - 1\right)^2\right]\right\}. \tag{19}$$

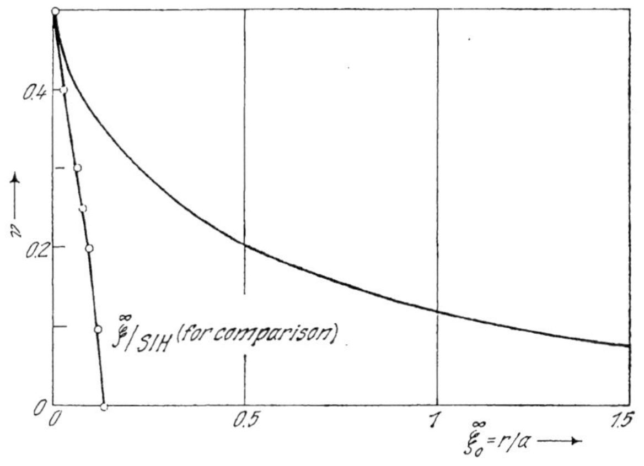

Fig. 4. Positions of initial points versus Poisson's ratio according the extended stress-parameter-criterion

We get maximum value of $\overset{\infty}{\Phi}$ according to equations (20):

$$\left\{\partial\overset{\infty}{\Phi}/\partial\xi = 0,\ \partial\overset{\infty}{\Phi}{}^2/\partial\xi^2 < 0\right\}_{|\xi=\overset{\infty}{\xi}_0}. \tag{20}$$

$\overset{\infty}{\xi}_0$ is the distance between the initial points of crack propagation and the crack tip. Fig. 4 shows the dependence of the distance upon Poisson's ratio ν.

4. Extension of the Multi-Axial-Value-Criterion

Kochendörfer and Scholl found in 1957 an inverse proportional connection between a so called value of multiaxiality of stress state and the fracture stress of steel samples by means of experimental methods, [13]. The multiaxial value is defined as follows:

$$\Omega \doteq 1 - \frac{\sqrt{3J_2}}{\sigma_1}. \tag{21}$$

J_2 is the 2nd invariant of stress deviator.

Now we interpret this parameter as a characteristic value of a crack fracture criterion, cf. [14]. The basic hypotheses of the simple criterion are the following:

1st hypothesis: Crack propagation proceeds in radial direction where $\overset{1}{\Omega}$ reaches a maximum, i.e.

$$\left\{\partial\overset{1}{\Omega}/\partial\vartheta = 0,\ \partial^2\overset{1}{\Omega}/\partial\vartheta^2 < 0\right\}_{|\vartheta=\overset{1}{\vartheta}_0}. \tag{22}$$

2nd hypothesis: The crack becomes unstable when $\overset{1}{K}_V(\overset{1}{\vartheta}_0)$ reaches a critical value K_{IC} where

$$\overset{1}{K}_V\left(\overset{1}{\vartheta}_0\right) \doteq K_I\left(\overset{1}{\Omega}_{\max}/\Omega_{\mathrm{Max}}\right)_{|\vartheta=\overset{1}{\vartheta}_0}. \tag{23}$$

If we extend this criterion and use the exact stress field solution around the crack tip of a Griffith-crack under Mode I-loading $(\overset{k}{\vartheta}_0 = 0^0)$, initial points are calculated too.

The distance between initial points and the crack tip reads as follows:

$$\overset{\infty}{\xi}_0 = \left\{1\Big/\left[1-\left(1-\frac{(\frac{1}{2}+\nu-\nu^2)}{(1-\nu+\nu^2)}\right)^2\right]\right\}^{1/2} - 1. \tag{24}$$

We take plane strain into account only. Fig. 5 shows the dependence of this distance upon Poisson's ratio.

A comparison of positions of initial points, calculated according to the three extended criteria, is shown in Table 2.

Table 2. Comparison of calculated positions of initial points (Griffith-crack, Mode I-loading, plane strain, $\vartheta_0 = 0^0$).

ν	$\overset{k}{\xi}_0$			
	Extended Sih-Criterion		Extended Stress-Parameter-Criterion	Extended Multi-Axial-Value-Criterion
	$k = 4$	$k = \infty$	$k = \infty$	$k = \infty$
0.0	0.137	0.122	(∞)	0.155
0.1	0.115	0.106	1.18	0.0684
0.2	0.0901	0.0853	0.512	0.0237
0.3	0.0613	0.0596	0.219	0.00517
0.4	0.0279	0.0276	0.0606	0.000346
(0,5)	(0)	(0)	(0)	(0)

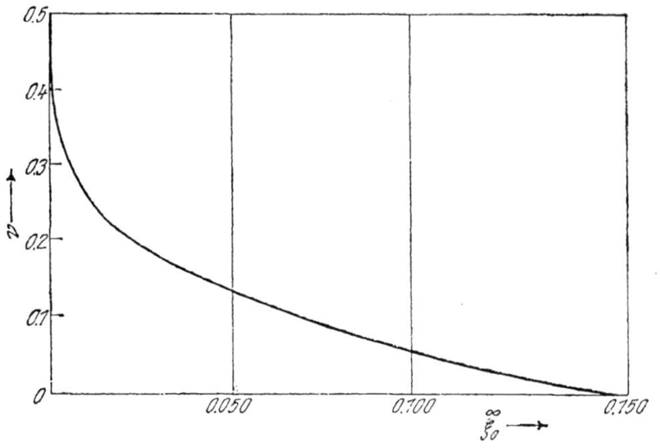

Fig. 5. Positions of initial points versus POISSON's ratio according the extended multi-axial-value-criterion

References

1 SIH, G. G., Mechanics of Fracture 1 — Methods of analysis and solutions of crack problems, Noordhoff Int. Publ., Leyden 1973.
2 FISCHER, K.-F., GÜNTHER, W., Gegenwärtiger Stand der Rißbruchmechanik im Hinblick auf eine praktische Nutzung, Maschinenbautechnik, 27 (1978) 2, S. 73—76.
3 FISCHER, K.-F., GÜNTHER, W., SCHMIDSBERGER, W., Zur Formulierung des Normalspannungskriteriums in der Rißbruchmechanik, Wiss. Beiträge der IH Zwickau, 4 (1978) 2, S. 119—125.
4 FISCHER, K.-F., GÜNTHER, W., Einbeziehung der Rißbruchmechanik in die theoretische Erklärung des Trennvorganges, Maschinenbautechnik, 28 (1979) 5, S. 220—223.
5 FISCHER, K.-F., Zur Ermittlung qulitativ neuer Bruchaussagen, die sich aus einer Erweiterung bekannter Bruchkriterien ergeben, Vorträge zum Problemseminar Bruchmechanik, Technische Universität Dresden 1979, Heft 1, S. 151—168.
6 COTTERELL, B., Notes on the path and stability of cracks, Int. Journal of Fracture Mechanics, 2 (1966), S. 526—533.
7 WILLIAMS, J. G., EWING, P. D., Fracture under complex stress — The angled crack problem, Int. Jour. of Fracture Mechanics, 8 (1972), S. 441—446.
8 SIH, G. C., Strain-energy-density-factor applied to mixed mode crack problems, Int. Journal of Fracture, 10 (1974) 3, S. 305—321.
9 FISCHER, K.-F., Zur Erweiterung des Bruchkriteriums von SIH, ZAMM, 58 (1978) 8, S. 331—336.
10 FISCHER, K.-F., On an extension of SIH's fracture criterion, Int. Journal of Fracture, 15 (1979) 1, S. R11—R14.
11 GARDE, A. M., WEISS, V., Brittle crack initiation at the elastic-plastic interface, Metallurgical Transactions, 3 (1972) 11, S. 2811 to 2817.
12 CHIANG, W. T., Fracture criteria for combined mode cracks, ICF 4, Fracture 1977 Vol. 4, Waterloo (Canada) 1977, S. 135—154.
13 KOCHENDÖRFER, A., SCHOLL, W., Die Sprödbruchneigung von Stählen in Abhänigkeit von Spannungszustand und Temperatur, Stahl und Eisen, 77 (1957), S. 1006—1018.
14 FISCHER, K.-F., Zum Verhalten der Mehrachsigkeitszahl in Rißspitzenumgebung (to be published).

Eingereicht am 20. 10. 1980

Anschrift: Dr.-Ing. K.-F. FISCHER, Ingenieurhochschule Zwickau, Abteilung Mathematik-Naturwissenschaften, Wissenschaftsbereich Technische Mechanik, DDR-9541 Zwickau, PSF 35

ZAMM 61, 235 – 241 (1981)

W. Splettstösser

On the Approximation of Random Processes by Convolution Processes

Glattheitseigenschaften von schwach stationären Prozessen und ihren Autokorrelationsfunktionen sind eng miteinander verknüpft. Ausgehend von diesem Zusammenhang lassen sich Aussagen über die Ordnung der Konvergenz gewisser Faltungsprozesse gegen gegebene stochastische Prozesse beweisen. Ist andererseits die Approximationsordnung bekannt, so kann man auf die Glattheit im Mittel der Prozesse schließen.

The smoothness in the mean of a weak sense stationary process is closely connected with that of its autocorrelation function. This enables one to prove results on the order of convergence of certain convolution processes to the given random process. On the other hand, it is possible to conclude smoothness properties, in case the order is known.

Свойства гладкости слабо стационарных случайных процессов и их автокорреляционных функций тесно связаны. Эта связь допускает доказывать результаты о порядке сходимости некоторых процессов свертывания на данные случайные процессы. Если, на другой стороне, известен порядок сходимости, потом можно делать выводы на гладкость в среднем процессов.

1. Indroduction

In physics as well as in most of the applied sciences deterministic functions are often supplied with random influences and, therefore, are dealt with as random functions or stochastic processes. Take the famous Langevin equation for a Brownian particle as an example, or a signal of speech, where the random portion of the function may be white noise or some other distortion in the transmission channel, perhaps given via a probability distribution. Then one often assumes that the statistics entering in that way does not depend "too closely" on the instants of time — a condition of stationarity. For example, the filtering problem of extracting a signal from noise in a best possible way can be solved for weak sense stationary (w.s.s.) processes by determining the Fourier transform of a convolution kernel (compare e.g. [1]).

The aim of this paper is to show how approximation theory, strictly speaking the approximation by singular convolution integrals, can be made available for w.s.s. processes. First of all, we prove some results on the continuity and differentiability of w.s.s. processes, which, to some extend, can be scraped together in textbooks, see e.g. [2]. Then we investigate a direct approximation theorem for continuous w.s.s. processes, and take as a concrete example the de La Vallée Poussin means. In this case the order of convergence is also evaluated for processes with Lipschitz-continuous derivatives. Furthermore, two inverse theorems from [3] are carried over, which may be applied to most of the known convolution integrals of Fejér's type.

Let us emphasize that only a few aspects of the large field of approximation theory are touched, although there are many results which are worthwhile to be "translated" into the language of random processes. There are a few topics, which we discuss briefly in a final outlook, namely saturation, periodic processes, and non-stationarity.

2. On continuity properties of random processes

2.1 Preliminaries

In all our considerations we take as a basis a probability space $(\Omega, \mathfrak{A}, \mathsf{P})$, where Ω is an arbitrary set, $\mathfrak{A}$ a σ-algebra on Ω, and P some probability measure on $\mathfrak{A}$, that is, $\mathsf{P}(\Omega) = 1$. A random process $X = X(t) = X(t, \omega)$ is then defined to be a real-(or complex-)valued function, measurable on the product space $\mathbb{R} \times \Omega$, $\mathbb{R}$ being the real line. For notational convenience we assume in everything that follows that all processes are real-valued. There is an obvious extension to complex-valued processes; it follows by interpreting $|X|$ as the complex modulus instead of the real one. Furthermore note that the t-domain need not necessarily be the whole line, finite subsets are also possible.

A process X belongs to $\mathfrak{L}^2$ if the expectation of $|X|^2$, i.e. $\mathsf{E}\{|X|^2\}$, is finite (finite energy process); the norm in the space $\mathfrak{L}^2$ is then defined by

$$||X(t)||_2 := \{\mathsf{E}\{|X(t)|^2\}\}^{1/2} := \left\{\int_\Omega |X(t, \omega)|^2 \, \mathrm{d}\mathsf{P}(\omega)\right\}^{1/2}; \tag{2.1}$$

it is finite for each $t \in \mathbb{R}$. In the sequel we always deal with *wide sense stationary* (w.s.s.) *processes*, meaning that the autocorrelation function of X, defined by

$$R_{X,X}(t, t + \tau) := \int_\Omega X(t, \omega) \, X(t + \tau, \omega) \, \mathrm{d}\mathsf{P}(\omega) , \qquad t, \tau \in \mathbb{R} , \tag{2.2}$$

is independent of $t \in \mathbb{R}$. In this case where the norm in (2.1) is also independent of $t \in \mathbb{R}$, we write $R_X(\tau) := R_{X,X}(t, t + \tau)$. It is well-known and easy to show that the autocorrelation function is even; it is maximal in modulus at $\tau = 0$, i.e., $|R_X(\tau)| \leq R_X(0)$ for each $\tau \in \mathbb{R}$.

2.2. Continuous processes

A random process $X \in \mathfrak{L}^2$ is said to be *continuous in the mean* (i.m.) *at* $t_0 \in \mathbb{R}$ if

$$\underset{h \to 0}{\text{l.i.m.}} \; \Delta_h X(t_0) = 0 ,$$

where the limit i.m. of the difference Δ_h of X with increment h tending to zero means that

$$\lim_{h\to 0} \left\{ \int_{\Omega} |X(t_0 + h, \omega) - X(t, \omega)|^2 \, dP(\omega) \right\}^{1/2} = 0 .$$

Similarly one defines *continuity i.m.* (on $\mathbb{R}$) and *uniform continuity i.m.* For w.s.s. processes a deep connection between continuity i.m. and continuity of the autocorrelation is known. This property is given in detail; in the following Lemma $C(\mathbb{R})$ denotes the class of uniform continuous and bounded functions on $\mathbb{R}$.

Lemma 2.1: *Let $X \in \mathfrak{L}^2$ be a w.s.s. process. The following conditions are equivalent:*

(i) *X is continuous i.m. at $t_0 \in \mathbb{R}$,*
(ii) *X is continuous i.m.* (on $\mathbb{R}$),
(iii) *X is uniformly continuous i.m.* (on $\mathbb{R}$),
(iv) *R_X is continuous at $\tau = 0$,*
(v) *R_X is continuous everywhere,*
(vi) *$R_X \in C(\mathbb{R})$.*

Proof: It suffices to show the implications (iv) $\Rightarrow$ (iii) and (i) $\Rightarrow$ (vi). Concerning (iv) $\Rightarrow$ (iii), the calculation of the norm of the first difference of X yields

$$||\Delta_h X(t)||_2^2 = 2R_X(0) - 2R_X(h) . \tag{2.3}$$

As the right-hand side of (2.3) tends to zero independently of $t \in \mathbb{R}$ in case that R_X is continuous in $\tau = 0$, uniform continuity i.m. of X is already shown. In order to prove that (i) $\Rightarrow$ (vi) one uses SCHWARZ's inequality to deduce for each $\tau \in \mathbb{R}$

$$|\Delta_h R_X(\tau)| = \left| \int_{\Omega} X(t_0 - \tau) \, [X(t_0 + h) - X(t_0)] \, dP \right| \leqq \{R_X(0)\}^{1/2} \, ||\Delta_h X(t_0)||_2 .$$

This already yields uniform continuity of R_X. Since R_X is uniformly bounded, as mentioned above, the proof is complete.

2.3 Differentiable processes

It is quite natural to say that a process $X \in \mathfrak{L}^2$ is *differentiable i.m. at $t \in \mathbb{R}$*, if there exists a process $\dot{X} \in \mathfrak{L}^2$ such that

$$\underset{h\to 0}{\text{l.i.m.}} \frac{1}{h} \Delta_h X(t) = \dot{X}(t) .$$

Differentiability on the whole line $\mathbb{R}$ as well as higher derivatives $X^{(r)}$ are defined in an analogous manner.

In order to give results for w.s.s. processes comparable to those of Lemma 1, it is useful to introduce the *second generalized derivative* $f^{[2]}$ for deterministic functions due to LOÈVE (compare e.g. [2, I]). It is defined by

$$f^{[2]}(t) = \lim_{h_1, h_2 \to 0} \frac{1}{h_1 h_2} \Delta_{h_2}(\Delta_{h_1} f)(t) , \tag{2.4}$$

where h_1 and h_2 must tend to zero independently. This is obviously a generalization of the usual second derivative; it can be seen by first letting h_1 und then h_2 tend to zero. Moreover, one has the following

Proposition 2.2: *If f'' exists as a continuous function on some interval (a, b), so does $f^{[2]}$, it being equal to f'' there.*

Proof: For any $t \in (a, b)$ and all h_1, h_2 with $[t - (|h_1| + |h_2|), t + |h_1| + |h_2|] \subset (a, b)$ the fundamental theorem of the calculus leads to

$$\Delta_{h_2}(\Delta_{h_1} f)(t) = \Delta_{h_2} \left(\int_t^{t+h_1} f'(u) \, du \right) = \int_t^{t+h_1} \left(\int_u^{u+h_2} f''(s) \, ds \right) du .$$

In view of the limit in (2.4) one has

$$\left| \frac{1}{h_1 h_2} \Delta_{h_2}(\Delta_{h_1} f)(t) - f''(t) \right| = \left| \frac{1}{h_1 h_2} \int_t^{t+h_1} \left(\int_u^{u+h_2} (f''(s) - f''(t)) \, ds \right) du \right| \leq \sup_{|h| \leq |h_1| + |h_2|} |f''(t + h) - f''(t)| .$$

The continuity of f'' at $t \in (a, b)$ implies that the last expression tends to zero for $h_1, h_2 \to 0$, and so the assertion follows immediately.

Now we are in the position to prove the result announced on the connection between differentiation of a w.s.s. process and its autocorrelation function. It can partially be found in text books in the field (see e. g. [4]), although sometimes with lax proofs [1].

Lemma 2.3: *Let $X \in \mathfrak{L}^2$ be a w.s.s. process. The following conditions are equivalent:*

(i) *X is differentiable i.m. at $t_0 \in \mathbb{R}$,*
(ii) *the first derivative $\dot{X}$ exists and is uniformly continuous i.m.* (on $\mathbb{R}$),
(iii) *$R_X^{[2]}$ exists at $\tau = 0$,*
(iv) *$R_X'' \in C(\mathbb{R})$.*

Proof: The step (ii) ⇒ (i) is trivial, and (iv) ⇒ (iii) is an easy consequence of Proposition 2.2. Concerning (iii) ⇒ (ii), the evaluation of

$$\left\| \frac{1}{h_1} \Delta_{h_1} X(t) - \frac{1}{h_2} \Delta_{h_2} X(t) \right\|_2^2 = \frac{1}{h_1^2} \Delta_{-h_1}(\Delta_{h_1} R_X)(0) + \frac{1}{h_2^2} \Delta_{-h_2}(\Delta_{h_2} R_X)(0) - \frac{2}{h_1 h_2} \Delta_{-h_1}(\Delta_{h_2} R_X)(0) \tag{2.5}$$

shows that the existence of $R_X^{[2]}(0)$ implies that $(\Delta_h X(t))/h$ is a CAUCHY sequence uniformly in $t \in \mathbb{R}$, and thus that $\dot{X}(t) \in \mathfrak{L}^2$ exists for each $t \in \mathbb{R}$. Moreover, the uniform convergence of $(\Delta_h X(t))/h$ together with the estimate

$$\|\dot{X}(t+h) - \dot{X}(t)\|_2 \leqq$$

$$\leqq \left\| \dot{X}(t+h) - \frac{1}{h} \Delta_h X(t+h) \right\|_2 + \left\| \frac{1}{2h} \Delta_{2h} X(t) - \frac{1}{h} \Delta_h X(t) \right\|_2 + \left\| \frac{1}{2h} \Delta_{2h} X(t) - \dot{X}(t) \right\|_2$$

yields the uniform continuity i.m. of $\dot{X}$. Note that we have shown the implication (i) ⇒ (ii) simultaneously, so that only the step (ii) ⇒ (iv) is left to complete the proof. To this end denote by

$$R_{X,Y}(t, t+\tau) := \int_\Omega X(t,\omega)\, Y(t+\tau,\omega)\, \mathrm{dP}(\omega)$$

the (cross-) correlation function of two processes X and Y defined on the same probability space $(\Omega, \mathfrak{A}, \mathrm{P})$. For the evaluation of R_X' we then have

$$\left| \frac{1}{h} \Delta_h R_X(\tau) - R_{X,\dot{X}}(t-\tau, t) \right| = \left| \int_\Omega X(t-\tau) \left(\frac{X(t+h) - X(t)}{h} - \dot{X}(t) \right) \mathrm{dP} \right| \leqq$$

$$\leqq \{R_X(0)\}^{1/2} \left\| \frac{1}{h} \Delta_h X(t) - \dot{X}(t) \right\|_2 . \tag{2.6}$$

Since the last term of (2.6) tends to zero independently of $t \in \mathbb{R}$, the first derivative $R_X'(\tau)$ of R_X exists and is equal to $R_{X,\dot{X}}(t-\tau, t) =: R_{X,\dot{X}}(\tau)$; this means that X and $\dot{X}$ are jointly w.s.s. In the same manner one arrives at the second derivative of R_X by

$$\left| \frac{1}{h} \Delta_h R_X'(\tau) + R_{\dot{X},\dot{X}}(t-\tau, t) \right| = \left| \int_\Omega \left(\frac{X(t-\tau-h) - X(t-\tau)}{h} + \dot{X}(t-\tau) \right) \dot{X}(t)\, \mathrm{dP} \right| \leqq$$

$$\leqq \left\| \frac{1}{-h} \Delta_{-h} X(t-\tau) - \dot{X}(t-\tau) \right\|_2 \|\dot{X}(t)\|_2 .$$

So the second derivative of $R_X(\tau)$ exists for each $\tau \in \mathbb{R}$ and is equal to the negative autocorrelation function of the w.s.s. process $\dot{X}$ which is continuous i.m. By Lemma 2.1 R_X'' then belongs to $C(\mathbb{R})$.

2.4 LIPSCHITZ classes

The last topic in this section is the characterization of processes which are continuous i.m. but not necessarily differentiable i.m. We therefore define the LIPSCHITZ *classes* $\mathrm{Lip}_L\,\alpha$ *of processes* $X \in \mathfrak{L}^2$ for some $0 < \alpha \leqq 1$ by

$$\mathrm{Lip}_L\,\alpha := \{X \in \mathfrak{L}^2; \omega(\mathfrak{L}^2; X; \delta) \leqq L\delta^\alpha\},$$

where $\omega(\mathfrak{L}^2; X; \delta) := \sup_{|h|<\delta} \|\Delta_h X(t)\|_2$ is a modulus of continuity i.m., and $L > 0$ the LIPSCHITZ constant; for generalizations compare [5], and [6] for periodic processes. Introducing the generalized LIPSCHITZ class $\mathrm{Lip}_M^*\,\beta$ of deterministic functions $f \in C(\mathbb{R})$ for $0 < \beta \leqq 2$ (compare e.g. [3]) by

$$\mathrm{Lip}_M^*\,\beta := \{f \in C(\mathbb{R}); \omega^*(C(\mathbb{R}); f; \delta) \leqq M\delta^\beta\},$$

where $\omega^*(C(\mathbb{R}); f; \delta) := \sup_{|h|<\delta} \|f(\cdot+h) - 2f(\cdot) + f(\cdot-h)\|$ is the classical generalized modulus of continuity, and $\|f\| = \|f(\cdot)\|$ denotes the usual sup-norm, one deduces

Lemma 2.4: *A w.s.s. process* $X \in \mathfrak{L}^2$ *belongs to* $\mathrm{Lip}_L\,\alpha$, *if and only if* R_X *belongs to* $\mathrm{Lip}_{L^2}^*\,2\alpha$.

Proof: First note that $\Delta_h X(t,\omega) \in \mathfrak{L}^2$ is a w.s.s. process itself because

$$R_{\Delta_h X}(t, t+\tau) = \int_\Omega (X(t+h) - X(t))\,(X(t+h+\tau) - X(t+\tau))\, \mathrm{dP} =$$

$$= 2R_X(\tau) - R_X(\tau+h) - R_X(\tau-h) \tag{2.7}$$

is independent of $t \in \mathbb{R}$. Being the autocorrelation of a w.s.s. process, $|R_{\Delta_h X}|$ takes its maximum at the origin, so that

$$\|R_X(\cdot+h) - 2R_X(\cdot) + R_X(\cdot-h)\| = 2R_X(0) - R_X(h) - R_X(-h) = \|\Delta_h X(t)\|_2^2 \tag{2.8}$$

for each $t \in \mathbb{R}$. Taking the sup on both sides of equation (2.8) leads to the desired result.

3. Approximation by convolution processes

As already mentioned, convolution of filtering is an important tool in signal theory for least-square estimation and prediction of noisy signals. It will be shown that convolution processes can be employed to approximate continuous processes in a similar way as singular integrals (e.g. of Fejér's type) approximate functions in the usual sense.

As is shown in [2, II], the convolution of a process $X \in \mathfrak{L}^2$ with a function g, which is absolutely integrable in Lebesgue's sense on $\mathbb{R}$ (in short $g \in L$), defined by

$$X * g(t, \omega) := \frac{1}{\sqrt{2\pi}} \int_{-\infty}^{\infty} X(u, \omega)\, g(t-u)\, du\,, \qquad t \in \mathbb{R}, \omega \in \Omega\,,$$

exists with probability 1 and belongs to $\mathfrak{L}^2$ itself. As is usual, the class L is provided with the norm

$$\|g\|_1 := \frac{1}{\sqrt{2\pi}} \int_{-\infty}^{\infty} |g(t)|\, dt\,.$$

Furthermore, if X is w.s.s., then X and $X * g$ are jointly w.s.s. since

$$R_{X*g,X}(t, t+\tau) = \int_{\Omega} \frac{1}{\sqrt{2\pi}} \int_{-\infty}^{\infty} X(u, \omega)\, g(t-u)\, du\, X(t+\tau, \omega)\, d\mathsf{P}(\omega) =$$

$$= \frac{1}{\sqrt{2\pi}} \int_{-\infty}^{\infty} R_X(t+\tau-u)\, g(t-u)\, du = R_X * g(\tau)\,; \tag{3.1}$$

here Fubini's theorem was used, and

$$f * g(\tau) = \frac{1}{\sqrt{2\pi}} \int_{-\infty}^{\infty} f(u)\, g(\tau-u)\, du\,, \qquad \tau \in \mathbb{R}$$

defines the usual convolution of two deterministic functions f and g. Similarly one deduces

$$R_{X,X*g}(t, t+\tau) = R * g(\tau)\,, \tag{3.2}$$

as well as

$$R_{X*g,X*g}(t, t+\tau) = R_X * g * g(\tau)\,, \tag{3.3}$$

showing the stationarity in the weak sense of the convolution process $X * g$.

A useful tool for results on approximation is the following estimate concerning the distance between X and $X * g$.

Proposition 3.1: *Let $X \in \mathfrak{L}^2$ be a w.s.s. process, and $g \in L$ an arbitrary function. Then*

$$\sup_{t \in \mathbb{R}} \|X(t) - X * g(t)\|_2 \leqq \{(1 + \|g\|_1)\, \|R_X - R_X * g\|\}^{1/2}\,.$$

Proof: Using the representations (3.1), (3.2) and (3.3) of the various correlation functions one deduces

$$R_{X-X*g,\,X-X*g}(t, t+\tau) = R_X(\tau) - 2R_X * g(\tau) + R_X * g * g(\tau)\,,$$

showing that the difference process $X - X * g$ is w.s.s. As the autocorrelation of a w.s.s. process is maximal in the origin, one derives for each $t \in \mathbb{R}$

$$\|X(t) - X * g(t)\|_2^2 = R_{X-X*g}(0) = \|R_X - 2R_X * g + R_X * g * g\| \leqq$$

$$\leqq \|R_X - R_X * g\| + \|(R_X * g - R_X) * g\| \leqq (1 + \|g\|_1)\, \|R_X - R_X * g\|\,; \tag{3.4}$$

the assertion now follows immediately.

We are now in the position to state a general result on the approximation of w.s.s. processes by certain convolution processes.

Theorem 3.2: *Let $X \in \mathfrak{L}^2$ be a w.s.s. process continuous i.m. at some $t_0 \in \mathbb{R}$. If the set of functions $\{g_\varrho\}_{\varrho>0} \subset L$ has the properties*

(i) $\int_{-\infty}^{\infty} g_\varrho(t)\, dt = \sqrt{2\pi}\,, \qquad \varrho > 0\,,$

(ii) $\|g_\varrho\|_1 \leqq M\,, \qquad \varrho > 0$, some $M > 0\,,$

(iii) $\lim_{\varrho \to \infty} \int_{\delta \leqq |t|} |g_\varrho(t)|\, dt = 0\,, \qquad \delta > 0\,,$

then

$$\lim_{\varrho \to \infty} \sup_{t \in \mathbb{R}} \|X(t) - X * g_\varrho(t)\|_2 = 0\,.$$

Proof: The continuity of X at $t_0 \in \mathbb{R}$ implies $R_X \in C(\mathbb{R})$ by Lemma 2.1. As the properties (i), (ii), (iii) are exactly those which are sufficient for $\{g_\varrho\}_{\varrho>0}$ to be an approximate identity, that is

$$\lim_{\varrho\to\infty} ||R_X - R_X * g_\varrho|| = 0$$

(compare e.g. [3]), the proof is complete with the help of Proposition 3.1.

Approximation theorems for harmonizable stochastic processes, which are in some sense related to our Theorem 3.2, can be found e.g. in [7], [8], where the Fourier-Stieltjes transform is used.

An important class of so-called convolution kernels $\{g_\varrho\}_{\varrho>0}$ are those of Fejér's type, which are defined by

$$g_\varrho(t) := \varrho g(\varrho t)\,, \qquad \varrho > 0,\, t \in \mathbb{R}, \tag{3.5}$$

where $g \in L$ is an arbitrary function satisfying

$$\int_{-\infty}^{\infty} g(t)\,dt = \sqrt{2\pi}\,. \tag{3.6}$$

For such kernels the properties (i), (ii), and (iii) of Theorem 3.2 are valid, and therefore

Corollary 3.3: *Let $X \in \mathfrak{L}^2$ be a w.s.s. process continuous i.m. at some $t_0 \in \mathbb{R}$. If $g \in L$ satisfies* (3.6), *then for g_ϱ defined by* (3.5) *one has*

$$\lim_{\varrho\to\infty} \sup_{t\in\mathbb{R}} ||X(t) - X * g_\varrho(t)||_2 = 0\,.$$

As smoothness conditions on X and R_X are closely related to each other, it is an easy task to prove a series of results concerning the order of approximation by specializing the convolution kernels and using Propositon 3.1. As an example, consider the means of de La Vallée Poussin, the kernel function $g = \theta$ of which is defined by

$$\theta(t) := \sqrt{\frac{2}{\pi}}\,\frac{\cos t - \cos 2t}{t^2}\,.$$

Corollary 3.4: *Let $X \in \mathfrak{L}^2$ be a w.s.s. process. If $X^{(r)}$ belongs to $\mathrm{Lip}_L\,\alpha$ for some $r \in \mathbb{P} := \{0, 1, 2, \ldots\}$, $\alpha \in (0, 1]$, then*

$$\sup_{t\in\mathbb{R}} ||X(t) - X * \theta_\varrho(t)||_2 \leqq c\varrho^{-r-\alpha}\,,$$

where the constant c is calculated for $\varrho \geqq 1$ to be

$$c = \left(\frac{4}{3} + \frac{2\sqrt{3}}{\pi}\right)\sqrt{\frac{\pi}{2}}\,L3^\alpha < 9,16L\,.$$

Proof: First note that $R_X^{(2r)}$ exists and belongs to $\mathrm{Lip}^*_{L^2}\,2\alpha$ because of Lemma 2.4. With an estimate in [9, p. 237] one can show similarly as in [10] that this implies for $\varrho \geqq 1$

$$||R_X - R_X * \theta_\varrho|| \leqq \left(\sqrt{3} + \frac{2\pi}{3}\right)3^{2\alpha}L^2\varrho^{-2r-2\alpha}\,.$$

By use of Proposition 3.1 one derives the desired estimate as well as the value of the constant.

Usually the rate of convergence investigated in Corollary 3.4 is written in O-notation as

$$\sup_{t\in\mathbb{R}} ||X(t) - X * \theta_\varrho(t)||_2 = O(\varrho^{-r-\alpha})\,, \qquad \varrho \to \infty\,.$$

4. Inverse results

In the previous section we indicated how to deduce a number of direct approximation theorems for random processes with the aid of Proposition 3.1 and Theorem 3.2. In a similar manner it is possible to translate inverse theorems from deterministic approximation theory into the language of stochastic processes. That is, we ask how smooth is a w.s.s. process i.m. provided it can be approximated by certain convolution integrals with a given rate of convergence. The main tool in this situation will be the equation (3.4). Our first result is

Theorem 4.1: *Let the kernel function $g \in C(\mathbb{R}) \cap L$ satisfy* (3.6) *and be twice differentiable with first and second derivative in $C(\mathbb{R}) \cap L$. If the w.s.s. process $X \in \mathfrak{L}^2$ is continuous i.m. at some $t_0 \in \mathbb{R}$ and can be approximated by $X * g_\varrho$, defined via* (3.5), *such that*

$$||X(t_1) - X * g_\varrho(t_1)||_2 = O(\varrho^{-\alpha})\,, \qquad \varrho \to \infty\,, \tag{4.1}$$

for some $t_1 \in \mathbb{R}$ and $\alpha > 0$, then

$$\omega(\mathfrak{L}^2; X; \delta) = \begin{cases} O(\delta^\alpha)\,, & 0 < \alpha < 1 \\ O(\delta\,|\log\delta|^{1/2})\,, & \alpha = 1\,, \qquad \delta \to 0+\,, \\ O(\delta)\,, & \alpha > 1 \end{cases} \tag{4.2}$$

uniformly in $t \in \mathbb{R}$.

Proof: With equation (3.4) the assumption (4.1) implies

$$||R_X - R_X * (2g_\varrho - g_\varrho * g_\varrho)|| = O(\varrho^{-2\alpha})\,, \qquad \varrho \to \infty\,, \tag{4.3}$$

R_X being in $C(\mathbb{R})$. The function $f(t) = 2g(t) - g * g(t)$ obviously belongs to $C(\mathbb{R}) \cap L$, satisfies (3.6), and admits derivatives up to the second order in $C(\mathbb{R}) \cap L$; moreover,

$$f_\varrho(t) = 2g_\varrho(t) - g_\varrho * g_\varrho(t) , \quad t \in \mathbb{R}, \varrho > 0 . \tag{4.3}$$

Therefore one can apply Theorem 3.5.1 in [3] to the function R_X and the kernel $\{f_\varrho\}_{\varrho>0}$, to deduce from

$$\omega^*(C(\mathbb{R}); R_X; \delta) = \begin{cases} O(\delta^\alpha) , & 0 < \alpha < 2 \\ O(\delta^2 |\log \delta|) , & \alpha = 2 , \quad \delta \to 0+ . \\ O(\delta^2) , & \alpha > 2 \end{cases} \tag{4.4}$$

The connection between the moduli $\omega(\mathfrak{L}^2; X; \delta)$ and $\omega^*(C(\mathbb{R}); R_X; \delta)$ given in Section 2.4 then leads to the desired result (4.2).

Let us state another somewhat deeper inverse theorem; it is an analogue of Theorem 3.5.2 in [3].

Theorem 4.2: *Let the kernel function $g \in C(\mathbb{R}) \cap L$ satisfy* (3.6) *and be* $(2r + 2)$*-times differentiable such that all derivatives up to order* $(2r + 2)$ *belong to $C(\mathbb{R}) \cap L$, r some positive integer. If the w.s.s. process $X \in \mathfrak{L}^2$ is continuous i.m. at some $t_0 \in \mathbb{R}$ and can be approximated by $X * g_\varrho$ such that*

$$||X(t_1) - X * g_\varrho(t_1)||_2 = O(\varrho^{-r-\alpha}) , \quad \varrho \to \infty ,$$

for some $t_1 \in \mathbb{R}$ and some $\alpha > 0$, then X is r-times differentiable i.m., all derivatives are uniformly continuous i.m., and

$$\omega(\mathfrak{L}^2; X^{(r)}; \delta) = \begin{cases} O(\delta^\alpha) , & 0 < \alpha < 1 \\ O(\delta |\log \delta|^{1/2}) , & \alpha = 1 , \quad \delta \to 0+ , \\ O(\delta) , & \alpha > 1 \end{cases}$$

uniformly in $t \in \mathbb{R}$.

Since many of the singular integrals known are based on kernel functions which are infinitely often differentiable, all derivatives being absolutely integrable, it is easy to find examples for the above theorems which are then valid for any positive integer r (cf. [3]).

5. Final remarks

In the preceding we established some special approximation theoretical results for random processes. On the other hand, the procedure used can be carried over to related topics in an obvious manner, so that our investigations might be exemplary. Three such problems are considered in this section.

The first is that of determining the optimal order of approximation of a random process by convolution processes when the kernel is fixed. Concerning the deterministic theory one asks whether there is a positive function $\varphi(\varrho)$ such that the order $O(\varphi(\varrho))$ is possible for a nontrivial class of functions, whereas only the invariant, e.g. constant, functions admit the order $o(\varphi(\varrho))$. Note that the example in Section 3 does not have this so-called saturation property, but the singular integral of Fejér, for example, does with $\varphi(\varrho) = 1/\varrho$. In case of w.s.s. processes the trivial processes X will be those with

$$R_X(\tau) = \mathsf{E}\{X(t)\, X(t + \tau)\} = \text{const.}, \quad t, \tau \in \mathbb{R};$$

this means a very strong stationarity.

Another problem is that of dealing with w.s.s. 2π-periodic processes, that is processes X with the property

$$||X(t + 2\pi) - X(t)||_2 = 0 , \quad t \in \mathbb{R} . \tag{5.1}$$

Using (2.7) and properties of w.s.s. processes it is not too difficult to show that (5.1) is valid if and only if the function R_X is of period 2π. This suggests one studies the theory of approximation of periodic w.s.s. processes by singular integrals similarly as in Sections 3 and 4.

In some applications the usual assumption of stationarity in the weak sense might be too restrictive. In this case one may start from the assumption that the norm (2.1) is bounded uniformly in $t \in \mathbb{R}$. No difficulties will arise when dealing with continuity and with differentiability, where partial derivatives of the autocorrelation $R_X(t, \tau)$ will appear. Problems do occur in connection with the modulus of continuity, if one wants to deduce the same orders of convergence in approximating X by singular integrals as in Section 3: the equation

$$||\Delta_h X(t)||_2^2 = R_X(t + h, t + h) - R_X(t, t + h) - (R_X(t + h, t) - R_X(t, t))$$

makes necessary the definition of a mixed modulus of continuity for deterministic functions.

This research was carried out under the DFG „Schwerpunktprogramm Digitale Signalverarbeitung" (Bu 166/35).

The author would like to thank Prof. Dr. P. L. Butzer for valuable comments as well as W. Engels for critical reading of the manuscript.

References

1 Papoulis, A., Probability, Random Variables, and Stochastic Processes, McGraw-Hill, New York 1965.
2 Blanc-Lapierre, A.; Fortet, R., Theory of Random Functions. I, II, Gordon and Breach, New York 1965, 1968.
3 Butzer, P. L.; Nessel, R. J., Fourier Analysis and Approximation. Vol. I: One-dimensional Theory, Birkhäuser Verlag, Basel-Academic Press, New York 1971.
4 Cramér, H.; Leadbetter, M. R., Stationary and Related Stochastic Processes, John Wiley & Sons, New York 1967.
5 Hahn, M. G., Conditions for sampling continuity and the central limit theorem, Ann. Probability 5 (1977), 351—360.
6 Kawata, T., Lipschitz classes of periodic stochastic processes, Keio Math. Sem. Rep. No. 3 (1978), 33—43.

7 ARIMOTO, A., Some theorems for harmonizable stochastic processes, Keio Math. Sem. Rep. No. **1** (1973), 24–30.
8 KAWATA, T., Some convergence theorems for stationary stochastic processes, Ann. Math. Statist. **30** (1959), 1192–1214.
9 ACHIESER, N. I., Vorlesungen über Approximationstheorie, Akademie Verlag, Berlin 1967.
10 SPLETTSTÖSSER, W.; STENS, R. L.; WILMES, G., On approximation by the interpolating series of G. VALIRON, Functiones et Approximatio **11** (1980), 39–56.

Eingereicht: 8. 1. 1980

Anschrift: Dr. W. SPLETTSTÖSSER, Lehrstuhl A für Mathematik, Rheinisch-Westfälische Technische Hochschule Aachen, Templergraben 55, D-5100 Aachen, BRD

ZAMM 61, 243 – 251 (1981)

D. Oelschlägel / H. Süsse

Optimierung mit Hilfe des Intervall-Newton-Verfahrens

Das bekannte Intervall-Newton-Verfahren wird so modifiziert, daß das Minimum einer streng konvexen Funktion über einem Intervall aus $I(R)$ eingeschlossen wird. Dabei werden auch eingangsbedingte Fehler berücksichtigt. Mit Hilfe dieses Verfahrens wird ein Algorithmus angegeben, der die gleiche Optimierungsaufgabe löst, wobei die Zielfunktion nicht konvex zu sein braucht. Weiterhin wird mit dem angegebenen Verfahren das mehrdimensionale Problem der Minimierung einer streng konvexen Funktion über einem Quader gelöst.

The well known interval Newton method is modified so that error bounds can be found for the minimum of a strictly convex function over an interval, which is element of $I(R)$. In this problem we have to consider data errors. With the help of this method an algorithm is developed, which solves the same problem with a nonconvex function. Furthermore the method can be extended to the multidimensional problem with a strictly convex function and a given interval vector.

Изменяется известный интервальный метод Ньютона так, что включается минимум строго выпуклой функции. При этом считаются и ошибки входа. С помощью этого метода строится алгоритм для решения самой задачи программирования, где целевая функция не обязательна выпукла. Кроме того при данном методе решается многоразмерная проблема минимизации строго выпуклой функции над интервальным вектором.

In der vorliegenden Arbeit sollen mit Hilfe des Intervall-Newton-Verfahrens einige spezielle Optimierungsaufgaben gelöst werden.

1. Bezeichnungen

Mit $A = [\underline{a}, \bar{a}]$, $\underline{a} \leqq \bar{a}$, $\underline{a}, \bar{a} \in R$, werden Intervalle aus $I(R)$ bezeichnet und mit $[x] = (X_i)$, $[x] \in V_n(I(R))$, Intervallvektoren. Weiterhin bezeichne $I(R_M)$ die Menge der Maschinenzahlen, und es gilt mit $[x] \in V_n(I(R))$, $[y] \in V_n(I(R_M))$

$$V_n([x]) := \{[z] \in V_n(I(R)) \mid [z] \subseteq [x]\}\,, \qquad \tilde{V}_n([y]) := \{[z] \in V_n(I(R_M)) \mid [z] \subseteq [y]\}\,.$$

Der Durchmesser $d(A)$ eines Intervalles A bzw. der Durchmesser $d([x])$ eines Intervallvektors $[x]$ sei durch

$$d(A) = \bar{a} - \underline{a}\,, \qquad d([x]) = \max_i d(X_i)$$

definiert. Der Betrag eines Intervalls sei $|A| = \max(|\underline{a}|, |\bar{a}|)$.

Es bezeichne F die stetige Intervallerweiterung einer stetigen Funktion f, deren Existenz stets vorausgesetzt sei. Im folgenden soll der Begriff „stetige Maschinen-Intervallerweiterung“ eingeführt werden. L bezeichne dabei die Anzahl der Ziffern der Mantisse einer Maschinenzahl sowie $x(L)$ eine Maschinenzahl. ($L \to \infty$ bedeute, daß auch die Länge des Exponenten „geeignet“ gegen ∞ strebt.)

Definition 1: Es sei $f \in C([x]^0)$ mit $[x]^0 \in V_n(I(R_M))$. $\tilde{F}_L$: $V_n([x]^0) \to I(R_M)$ heißt dann *stetige Maschinen-Intervallerweiterung* von f im Intervallvektor $[x]^0$, falls die stetige Intervallerweiterung F zu f in $[x]^0$ existiert und wenn gilt:

a) $\{f(x) \mid x \in [x]\} \subseteq F([x]) \subseteq \tilde{F}_L([x])$ für alle $[x] \in \tilde{V}_n([x]^0)$.

b) $\tilde{F}_L$ ist inklusionsmonoton über $\tilde{V}_n([x]^0)$.

c) Mit $[x] \in V_n([x]^0)$, $[x] \subseteq [\underline{x}(L), \bar{x}(L)] \in \tilde{V}_n([x]^0)$, $L \geqq 1$, folgt aus $\lim\limits_{L\to\infty} [\underline{x}(L), \bar{x}(L)] = [x]$ die Beziehung $\lim\limits_{L\to\infty} \tilde{F}_L([\underline{x}(L), \bar{x}(L)]) = F([x])$.

Die stetige Maschinen-Intervallerweiterung ist somit an eine bestimmte stetige Intervallerweiterung gebunden, ist ihr somit zugeordnet.

Ist $* \in \{+, -, \cdot, :\}$ eine Grundoperation der Intervallarithmetik, dann bezeichne $\circledast$ die entsprechende Operation der Maschinen-Intervallarithmetik sowie $\boxed{*}$ die entsprechende Operation der „gewöhnlichen“ Maschinen-Arithmetik.

Die Grundlagen der Intervall-Mathematik werden im folgenden als bekannt vorausgesetzt, man vgl. hierzu Alefeld und Herzberger [1].

2. Modifizierte Newton-Verfahren

Gegeben sei die Optimierungsaufgabe

$$\hat{z} = f(\hat{x}) = \min f(x) \quad \text{bei} \quad x \in A \tag{1}$$

mit $A = [\underline{a}, \bar{a}] \in I(R)$ und $f\colon D \to R^1$, $A \subset D \subset R^1$, f sei über A streng konvex. Die Existenz und Eindeutigkeit von $\hat{x}$ sind somit gesichert.

Wir stellen folgende numerische Aufgabe:

Man finde Intervalle X und Z aus $I(R)$ mit „möglichst kleinen“ Durchmessern, die die Relationen

$$\text{a) } \hat{x} \in X \quad \text{und} \quad \text{b) } \hat{z} \in Z$$

erfüllen.

Diese Aufgabe wurde in der Literatur von DUSSEL [3] und BROSSART [2] mit den Methoden der Intervall-Mathematik gelöst. DUSSEL benutzt ein einfaches Halbierungsverfahren, während BROSSART ein ableitungsfreies Verfahren vorschlägt. Beide Verfahren erzeugen eine nur linear konvergente Intervallfolge. In der vorliegenden Arbeit soll diese Aufgabe mit dem bekannten Intervall-NEWTON-Verfahren gelöst werden.

Folgende Vorgehensweise zur Berechnung von $\hat{x}$ ist zunächst naheliegend (dabei wird $f(x)$ als zweimal stetig differenzierbar über D angenommen):

Man berechne $f'(\underline{a})$, $f'(\bar{a})$ und betrachte die folgende Fallunterscheidung:

— $\operatorname{Sign} f'(\underline{a}) = \operatorname{sign} f'(\bar{a}) = +1 \Rightarrow \hat{x} = \underline{a}$.

— $\operatorname{Sign} f'(\underline{a}) = \operatorname{sign} f'(\bar{a}) = -1 \Rightarrow \hat{x} = \bar{a}$.

— $f'(\underline{a}) = 0 \vee f'(\bar{a}) = 0 \Rightarrow \hat{x} = \underline{a} \vee \hat{x} = \bar{a}$.

— $\operatorname{Sign} f'(\underline{a}) \neq \operatorname{sign} f'(\bar{a}) \Rightarrow \exists\, \xi \in A\colon \hat{x} = \xi \wedge f'(\xi) = 0$.

Im letzten Falle schließe man ξ mittels eines Intervall-Nullstellenverfahrens ein.

Realisiert man diese Fallunterscheidungen auf einer EDVA und benutzt eine Maschinenintervallarithmetik, dann werden statt der exakten Funktionswerte Intervalle berechnet. Enthalten diese aber die Null, so versagt obiger Algorithmus. Bei dem folgenden modifizierten NEWTON-Verfahren brauchen keine Fallunterscheidungen durchgeführt werden. Fallunterscheidungen brauchen auch dann nicht durchgeführt werden, wenn das folgende Verfahren in $I(R_M)$ übertragen wird, was im übernächsten Abschnitt gezeigt wird.

Satz 1: *Zusätzlich zu den genannten Voraussetzungen mögen Zahlen* $\underline{m}, \bar{m} \in R$ *existieren mit*

$$0 < \underline{m} \leqq f''(x) \leqq \bar{m} \quad \text{für alle} \quad x \in A. \tag{2}$$

Beginnend mit $X^0 = [\underline{a}, \bar{a}]$ *erzeugt das Iterationsverfahren*

$$M^k = F''(X^k) \cap [\underline{m}, \bar{m}], \qquad Y^{k+1} = x^k - \frac{f'(x^k)}{M^k}, \qquad Z^{k+1} = Y^{k+1} \cap X^k,$$

$$X^{k+1} = \begin{cases} Z^{k+1}, & \text{falls } Z^{k+1} \neq \emptyset \\ \underline{x}^k, & \text{falls } \bar{y}^{k+1} < \underline{x}^k, \\ \bar{x}^k, & \text{falls } \underline{y}^{k+1} > \bar{x}^k \end{cases} \tag{3}$$

$$x^k \in X^k, \qquad k = 0, 1, 2, \dots,$$

eine Intervallfolge $\{X^k\}$. *Das Verfahren wird mit* X^{k_0+1} *abgebrochen, falls* $f'(x^{k_0}) = 0$ *oder* $Z^{k_0+1} = \emptyset$ *ist. Für die nach* (3) *erzeugte Folge* $\{X^k\}$ *gilt:*

(B1) $\hat{x} \in X^k$, $k \geqq 0$;

(B2) $\{X^k\}$ *ist monoton, d. h.* $X^0 \supseteq X^1 \supseteq X^2 \supseteq \dots$;

(B3) *ist* $\bar{y}^{k_0+1} < \underline{x}^{k_0}$, *dann folgt* $\hat{x} = \underline{x}^{k_0} = \underline{a}$,
ist $\underline{y}^{k_0+1} > \bar{x}$, *dann folgt* $\hat{x} = \bar{x}^{k_0} = \bar{a}$;

(B4) $d(X^{k+1}) \leqq \left(1 - \dfrac{\underline{m}^k}{\bar{m}^k}\right) \cdot d(X^k)$, $\quad k \geqq 0$.

Erfüllt $F''(X)$ *die Bedingung*

$$d(F''(X)) \leqq c \cdot d(X), \qquad c \geqq 0, \qquad X \subseteq A, \tag{4}$$

dann gilt sogar

$$d(X^{k+1}) \leqq K \cdot d(X^k)^2, \qquad K \geqq 0;$$

(B5) $\lim\limits_{k \to \infty} X^k = \hat{x}$;

(B6) *ist entweder* $f'(x) < 0$ *oder* $f'(x) > 0$ *für alle* $x \in A$, *dann bricht die Iteration* (3) *nach endlich vielen Schritten mit dem Abbrechindex* k_0 *ab.*

Beweis:

Zu (B1): Existiert ein $\xi \in [\underline{a}, \bar{a}]$ mit $f'(\xi) = 0$, so ist $\hat{x} = \xi$, und das Verfahren (3) geht in das bekannte Intervall-NEWTON-Verfahren über, so daß $\hat{x} = \xi \in X^k$, $k \geqq 0$ gesichert ist.

Man betrachte nun den Fall, daß $f'(x) > 0$ für alle $x \in A$ ist, dann folgt $\hat{x} = \underline{a}$. Weiterhin sei $\hat{x} \in X^k$. Da $\underline{m}^k > 0$, folgt mit $Y^{k+1} = x^k - \dfrac{f'(x^k)}{M^k}$ sofort die Beziehung $\bar{y}^{k+1} < x^k$. Nun muß noch die Beziehung $y^{k+1} \leqq \underline{x}^k$ gezeigt werden, denn dann folgt die zu beweisende Relation $\hat{x} \in X^{k+1}$.

Nach einem Satz von NICKEL [10] folgt aus der Beziehung $N(x^k, X^k) = x^k - \dfrac{f'(x^k)}{M^k} \subseteq X^k$ die Existenz einer Nullstelle ξ in X^k mit $f'(\xi) = 0$. Auf Grund obiger Voraussetzung existiert aber keine Nullstelle in X^k, so daß mit $\bar{y}^{k+1} < x^k$ nur $\underline{y}^{k+1} < \underline{x}^k$ gelten kann.

Durch vollständige Induktion folgt somit $\hat{x} = \underline{a} \in X^k$ für $k \geqq 0$.

Der Fall $f'(x) < 0$ für alle $x \in A$ kann völlig analog bewiesen werden.

Zu (B3): Folgt aus dem Beweis zu (B1).

Die Behauptungen (B2), (B4), (B5) und (B6) gelten völlig analog wie beim Intervall-NEWTON-Verfahren (siehe ALEFELD und HERZBERGER [1]).

Analog wie bei Verfahren zur Nullstellenbestimmung (man vgl. HERZBERGER [7], OELSCHLÄGEL und SÜSSE [12]) kann durch das bekannte Halbierungsprinzip eine Konvergenzbeschleunigung im Sinne der Inkulsion erreicht werden. Das Wesentliche dabei ist, daß der Rechenaufwand derselbe ist. Mit $x^k = m(X^k) = \frac{\underline{x}^k + \overline{x}^k}{2}$ geht dann (3) über in die Iterationsvorschrift:

$$U^k = \begin{cases} [\underline{x}^k, m(X^k)], & \text{falls } f'(m(X^k)) > 0 \\ m(X^k), & \text{falls } f'(m(X^k)) = 0, \\ [m(X^k), \overline{x}^k], & \text{falls } f'(m(X^k)) < 0 \end{cases}$$

$$M^k = F''(U^k) \cap [\underline{m}, \overline{m}], \qquad Y^{k+1} = m(X^k) - \frac{f'(m(X^k))}{M^k}, \qquad Z^{k+1} = Y^{k+1} \cap U^k,$$

$$X^{k+1} = \begin{cases} Z^{k+1}, & \text{falls } Z^{k+1} \neq \emptyset \\ \underline{u}^k, & \text{falls } \overline{y}^{k+1} < \underline{u}^k, \quad k = 0, 1, 2, \ldots \\ \overline{u}^k, & \text{falls } \underline{y}^{k+1} > \overline{u}^k \end{cases} \tag{5}$$

Kann a priori $\underline{m} > 0$ nicht gesichert werden, dann wird auf Grund der Konvexität von f die Größe $\underline{m} = 0$ gesetzt.

Ist $\underline{m}^k = 0$, dann geht die Iterationsfolge (5) über in

$$y^{k+1} = m(X^k) - \frac{f'(m(X^k))}{\overline{m}^k}, \qquad \overline{m}^k > 0,$$

$$X^{k+1} = \begin{cases} \underline{u}^k, & \text{falls } y^{k+1} < \underline{u}^k \\ \overline{u}^k, & \text{falls } y^{k+1} > \overline{u}^k. \\ Z^{k+1} & \text{sonst} \end{cases} \tag{6}$$

Ist $\underline{u}^k \leqq y^{k+1} \leqq \overline{u}^k$, dann ergibt sich

$$Z^{k+1} = \begin{cases} [\underline{u}^k, y^{k+1}], & \text{falls } f'(m(X^k)) > 0, \\ [y^{k+1}, \overline{u}^k], & \text{falls } f'(m(X^k)) < 0. \end{cases} \quad k = 0, 1, 2, \ldots$$

Tritt eine der Bedingungen $f'(m(X^k)) = 0$, $y^{k_0+1} > \underline{u}^{k_0}$, $y^{k_0+1} < \overline{u}^{k_0}$ ein, dann wird die Iteration abgebrochen. Existiert ein k_1 mit $\underline{m}^{k_1} > 0$, dann wird ab k_1 statt (6) wieder die Iteration (5) verwendet.

Für die nach (6) *erzeugte Folge* $\{X^k\}$ *gilt:*

(B1) $\hat{x} \in X^k$, $k \geqq 0$,

(B2) $\{X^k\}$ *ist monoton.*

(B3) *Ist* $y^{k_0+1} < \underline{u}^{k_0}$, *dann folgt* $\hat{x} = \underline{u}^{k_0} = \underline{a}$,
Ist $y^{k_0+1} > \overline{u}^{k_0}$, *dann folgt* $\hat{x} = \overline{u}^{k_0} = \overline{a}$;

(B4) $d(X^{k+1}) < \frac{1}{2} d(X^k)$, $\quad k \geqq 0$;

(B5) $\lim_{k \to \infty} X^k = \hat{x}$;

(B6) *ist* $f'(x) > 0$ *oder* $f'(x) < 0$ *für alle* $x \in A$, *dann bricht die Iteration nach endlich vielen Schritten ab.*

Der Beweis ist leicht zu führen. Insbesondere ist (6) dann zu verwenden, wenn $\hat{x}$ eine mehrfache Nullstelle von $f'(x)$ ist, wobei die Folge nach (6) im Sinne der Inklusion immer noch schneller konvergiert als das gewöhnliche Halbierungsverfahren.

3. Unscharfe Eingangsdaten

Die Aufgabe (1) soll nun so erweitert werden, daß auch eingangsbedingte Fehler, z. B. Meßfehler physikalischer Größen, berücksichtigt werden. Dazu werden Parameter c_i, $i = 1, \ldots, n$, eingeführt, die unabhängig voneinander in gegebenen Intervallen C_i aus $I(R)$ variieren mögen. Mit $c = (c_i)$, $[c] = (C_i)$ geht dann (1) in die Aufgabe

$$\hat{z}(c) = f(\hat{x}(c); c) = \min f(x; c), \qquad c \in [c] \quad \text{bei} \quad x \in A \tag{7}$$

über. Mit $\hat{X} = \{\hat{x}(c) \in R^1 \mid c \in [c]\}$ werde die *Menge der Minimalpunkte* und mit $\hat{Z} = \{\hat{z}(c) \in R^1 \mid c \in [c]\}$ die *Menge der Optimalwerte* bezeichnet.

Gesucht sind „günstige" Oberhüllen bzw. die Hüllen der Mengen $\hat{X}$ und $\hat{Z}$. In der Literatur wurde diese Aufgabenstellung bisher noch nicht behandelt. Für beliebiges $c \in [c]$ seien die gleichen Forderungen an $f(x; c)$ wie in Satz 1 gestellt, d. h. es gelte z. B.

$$0 < \underline{m} \leqq f''(x; c) \leqq \overline{m}, \quad \text{für alle} \quad x \in A, \quad \text{und alle} \quad c \in [c].$$

Zusätzlich seien f, f' und f'' über $A \times [c]$ stetig.

Satz 2: *Beginnend mit $X^0 = A$ erzeugt das Iterationsverfahren*

$$U^k = \begin{cases} [\underline{x}^k, m(X^k)], & \text{falls} \quad \inf F'(m(X^k); [c]) > 0 \\ X^k \quad , & \text{falls} \quad 0 \in F'(m(X^k); [c]) \quad , \\ [m(X^k), \bar{x}^k], & \text{falls} \quad \sup F'(m(X^k); [c]) < 0 \end{cases}$$

$$M^k = F''(U^k; [c]) \cap [\underline{m}, \overline{m}] , \qquad Y^{k+1} = m(X^k) - \frac{F'(m(X^k); [c])}{M^k} , \qquad Z^{k+1} = Y^{k+1} \cap U^k ,$$

$$X^{k+1} = \begin{cases} Z^{k+1}, & \text{falls} \quad Z^{k+1} \neq \emptyset \\ \underline{u}^k \quad , & \text{falls} \quad \bar{y}^{k+1} < \underline{u}^k , \\ \bar{u}^k \quad , & \text{falls} \quad \underline{y}^{k+1} > \bar{u}^k \end{cases} \qquad k = 0, 1, 2, \ldots , \tag{8}$$

eine Intervallfolge $\{X^k\}$. *Abgebrochen wird, falls* $Z^{k_0+1} = \emptyset$ *ist. Für die nach* (8) *erzeugte Folge* $\{X^k\}$ *gilt:*

(B1) $\hat{X} \subseteq X^k, k \geqq 0,$

(B2) $\{X^k\}$ *ist monoton;*

(B3) *ist* $\bar{y}^{k_0+1} > \underline{u}^{k_0}$, *so folgt* $\hat{X} = \underline{a}$,
ist $\underline{y}^{k_0+1} < \bar{u}^{k_0}$, *so folgt* $\hat{X} = \bar{a}$;

(B4) $\lim_{k\to\infty} X^k = X^*$ *mit* $X^* \subseteq \hat{X}$, *wobei* X^* *i. a. nur Oberhülle von* $\hat{X}$ *ist.*

(B5) *Ist* $\inf F'(x; [c]) > 0$ *oder* $\sup F'(x; [c]) < 0$ *für alle* $x \in A$, *dann bricht die Iteration nach endlich vielen Schritten mit dem Abbrechindex* k_0 *ab.*

Der Beweis folgt aus dem Beweis zu Satz 1.

Die Menge $\hat{Z}$ wird dann durch $Z = F(X^*; [c])$ eingeschlossen.

Wann können nun die Hüllen für $\hat{X}$ und $\hat{Z}$ berechnet werden? Auskunft geben die folgenden beiden Sätze.

Satz 3: *Es seien* $c^1, c^2 \in [c]$ *fest gewählt. Ist die Monotoniebedingung*

$$f(x; c^1) \leqq f(x; c) \leqq f(x; c^2) \quad \text{für alle} \quad x \in A \quad \text{und alle} \quad c \in [c]$$

erfüllt, dann ist $Z^H = [\hat{z}(c^1), \hat{z}(c^2)]$ *Hülle von* $\hat{Z}$.

Satz 4: *Es seien* $c^{11}, c^{22} \in [c]$ *fest gewählt. Ist die Monotoniebedingung*

$$f'(x, c^{11}) \leqq f'(x; c) \leqq f'(x; c^{22}) \quad \text{für alle} \quad x \in A \quad \text{und alle} \quad c \in [c]$$

erfüllt, dann ist $X^H = [\hat{x}(c^{22}), \hat{x}(c^{11})]$ *Hülle von* $\hat{X}$.

Die Behauptungen beider Sätze sind geometrisch unmittelbar einzusehen. Unter den Voraussetzungen der Sätze 3 und 4 können somit die Hüllen berechnet werden, indem die nichtparametrischen Teilprobleme mit dem Algorithmus (5) bzw. (6) gelöst werden. Dabei können maximal 4 Teilprobleme auftreten. Sind die Voraussetzungen der Sätze 3 und 4 nicht erfüllt, dann wird mit Hilfe des Verfahrens (8) eine Oberhülle von $\hat{X}$ berechnet, diese sei mit X^* bezeichnet. Diese Oberhülle X^* kann folgendermaßen verbessert werden.

Es gelte z. B. $\underline{a} < \underline{x}^* < \bar{x}^* \leqq \bar{a}$. *Man führe zur Verbesserung von* $\underline{x}^*$ *den Algorithmus* (8) *n-mal mit den Startintervallen*

$$X_i^0 = \left[\underline{a}, \underline{x}^* + \frac{i}{n} d(X^*)\right], \qquad i = 1(1)\, n , \tag{9}$$

durch, wobei $n > 1$ *unter Berücksichtigung des dadurch entstehenden Mehraufwandes geeignet zu wählen ist. Das Grenzelement der Folge mit dem Startintervall* X_i^0 *sei mit* X_i^*, $i = 1(1)\, n$ *bezeichnet.*

Satz 5: *Unter obigen Voraussetzungen folgen mit* $\underline{x}^{**} = \max_i \underline{x}_i^*$ *die Behauptungen:*

(B1) $\underline{x}^{**} \geqq \underline{x}^*$,

(B2) $[\underline{x}^{**}, \underline{x}^*] \supseteq \hat{X}$.

Beweis:

Zu (B1): Wegen $X_i^0 \subseteq A$ folgt (B1) aus der Teilmengeneigenschaft sowie aus der Ungleichung $\bar{x}_i^0 \geqq \underline{x}^*$.

Zu (B2): Wegen $\underline{a} < \underline{x}^* < \bar{a}$ ist das Minimum $\hat{x}_i(c)$ von $f(x; c)$ über X_i^0 entweder freies Minimum, oder es ist gleich $\bar{x}_i^0$. Ist $\hat{x}_i(c)$ freies Minimum, dann erhält man mit Verfahren (8) eine Einschließung von $\hat{x}_i(c)$. Ist $\hat{x}_i(c) = \bar{x}_i^0$, dann kann wegen der Konvexität von f der Minimalpunkt $\hat{x}(c)$ von $f(x; c)$ über A nur „rechts" von $\bar{x}_i^0$ liegen. Da diese Überlegungen für $i = 1(1)\, n$ richtig sind, folgt schließlich die Behauptung.

Die Größe $\underline{x}^{**}$ ist oft eine wesentlich günstigere Schranke als $\underline{x}^*$. Zur Verbesserung von $\bar{x}^{**}$ gehe man völlig analog vor.

4. Realisierung auf einer EDVA

Da sich die maschinenmäßige Durchführung von Intervallalgorithmen und die daraus abgeleiteten Aussagen von der theoretischen Durchführung und den dazugehörigen Behauptungen unterscheiden, muß jeder Intervallalgorithmus genaugenommen noch einmal „maschinenmäßig" formuliert werden. In der Literatur wird das oft nur mit der

Bemerkung abgetan, daß eine geeignete Maschinen-Intervallarithmetik zu verwenden sei. Bei der maschinenmäßigen Realisierung der Algorithmen (3), (5), (6) und (8) treten allerdings einige Schwierigkeiten auf. Die Intervallgrenzen $\underline{a}$ und $\overline{a}$ müssen keine Maschinenzahlen sein, es können Darstellungsfehler oder Konvertierungsfehler auftreten. Nun ist es nicht einfach möglich, ein Maschinenintervall A_M mit $A_M \supseteq A$ zu wählen, da der Minimalpunkt über A nicht der gleiche wie über A_M zu sein braucht. Deshalb werden auf einer EDVA zwei Maschinenintervalle $A_1, A_2 \in I(R_M)$ mit $\underline{a} \in A_1$, $\overline{a} \in A_2$ und „möglichst kleinen" Durchmessern bestimmt, um eine maschinenmäßige Durchführung der Algorithmen zu ermöglichen. Im folgenden wird die maschinenmäßige Durchführung des Verfahrens (5) angegeben, alle anderen Verfahren können analog in $I(R_M)$ übertragen werden.

Satz 6: *Beginnend mit $X^0 = [\underline{a}_1, \overline{a}_2] \in I(R_M)$ erzeugt das Iterationsverfahren*

$$U^k = \begin{cases} [\underline{x}^k, m(X^k)], & \text{falls} \quad \inf \tilde{F}'_L(m(X^k)) > 0 \\ X^k \quad , & \text{falls} \quad 0 \in \tilde{F}'_L(m(X^k)) \\ [m(X^k), \overline{x}^k], & \text{falls} \quad \sup \tilde{F}'_L(m(X^k)) < 0 \end{cases},$$

$$M^k = \tilde{F}''_L(U^k) \cap [\underline{m}, \overline{m}], \qquad \underline{m} > 0, \qquad \underline{m}, \overline{m} \in R_M,$$

$$Y^{k+1} = m(X^k) \ominus \tilde{F}'_L(\tilde{m}(X^k)) \odot M^k, \qquad Z^{k+1} = Y^{k+1} \cap U^k,$$

$$X^{k+1} = \begin{cases} A_1, & \text{falls} \quad \overline{y}^{k+1} < \overline{a}_1 \\ A_2, & \text{falls} \quad \underline{y}^{k+1} > \underline{a}_2 \\ Z^{k+1} & \text{sonst} \end{cases} \tag{5'}$$

mit $\tilde{m}(X^k) = (\underline{x}^k \boxed{+} \overline{x}^k) \boxed{:} 2$, $k = 0, 1, 2, \ldots$, eine Intervallfolge $\{X^k\}$ mit $X^k \in I(R_M)$, $k \geqq 0$. Abgebrochen wird die Iteration mit dem Abbrechindex k_1, wenn zuerst eine der folgenden Bedingungen eintrifft:

a) $\tilde{m}(X^{k_1+1}) \notin X^{k_1+1}$;

b) $X^0 \supset X^1 \supset \ldots \supset X^{k_1} = X^{k_1+1}$.

Für die nach (5) *erzeugte, endliche Folge $\{X^k\}$ gilt:*

(B1) $\hat{x} \in X^k$, $k = 0, 1, \ldots, k_1$;

(B2) $X^0 \supset X^1 \supset \ldots \supset X^{k_1}$.

Beweis: Auf Grund der Inklusionsmonotonie von ⊛ und der Eigenschaften einer stetigen Maschinen-Intervallerweiterung, der Durchschnittsbildung und der Definition des Abbrechindexes folgen sofort die Behauptungen.

Abschätzungen der Form $d(X^{k+1}) \leqq \frac{1}{2}\left(1 - \frac{\underline{m}}{\overline{m}}\right) d(X^k)$ für $k = 0, 1, \ldots, k_1$ gelten nicht mehr, da nicht notwendig $d(\tilde{F}'_L(\tilde{m}(X^k))) = 0$ und $\tilde{m}(X^k) = m(X^k)$ ist.

Im Gegensatz zu den üblichen Nullstellenalgorithmen der Intervall-Mathematik kann der Algorithmus (5') Punktintervalle $X^{k_1} \in R_M$ erzeugen, nämlich dann, wenn $\underline{a}_1 = \overline{a}_1 \in R_M$, $\underline{a}_2 = \overline{a}_2 \in R_M$ ist und eine der Bedingungen $\overline{y}^{k_1+1} < \overline{a}_1$, $\underline{y}^{k_1+1} > \underline{a}_2$ eintrifft. Gewisse Schwierigkeiten bereitet die maschinenmäßige Durchführung des Verfahrens (6). Aus den Entscheidungen $f'(m(X^k)) > 0$, $f'(m(X^k)) = 0$, $f'(m(X^k)) < 0$ werden maschinenmäßig die Fallunterscheidungen

$$\inf \tilde{F}'_L(\tilde{m}(X^k)) > 0, \qquad 0 \in \tilde{F}'_L(\tilde{m}(X^k)), \qquad \sup \tilde{F}'_L(\tilde{m}(X^k)) < 0.$$

Im Falle $0 \in \tilde{F}'_L(\tilde{m}(X^k))$ ist keine Aussage über die Lage des Optimums möglich. Folglich müssen $\tilde{m}(X^k)$ „benachbarte" Maschinenzahlen $x^k \in R_M$ so gewählt werden, daß $0 \notin \tilde{F}'_L(x^k)$ gilt. Ist dies nicht möglich, dann muß abgebrochen werden.

5. Nichtkonvexe Zielfunktion

Ähnlich (1) wird die Aufgabe

$$\hat{z} = f(\hat{x}) = \min f(x) \quad \text{bei} \quad x \in A \tag{10}$$

betrachtet. Dabei sei $f\colon D \to R^1$ mit $A \subset D \subset R^1$, $f(x) \in C^2(A)$. Weiterhin sei $f(x)$ über A nirgends stückweise konstant und besitze höchstens endlich viele stationäre Stellen. Die Aufgabe (10) besitzt folglich endlich viele, voneinander isolierte Minimalpunkte. Die numerische Aufgabenstellung soll wiederum darin bestehen, endlich viele Intervalle anzugeben, die die Minimalpunkte $\hat{x}$ und $\hat{z}$ „möglichst gut" einschließen.

Von MANCINI und WILDE [9] wurde die Aufgabe (10) so gelöst, indem alle Extremwerte von $f(x)$ über A eingeschlossen wurden. Weiterhin wurde die Aufgabe (10) von HANSEN [5] mit Hilfe der Intervall-Mathematik gelöst. HANSEN entwickelte einen Algorithmus, der eine Kombination zwischen einem von HANSEN [4] angegebenen erweiterten Intervall-NEWTON-Verfahren zur Nullstellenbestimmung und zu einfachen Funktionswertvergleichen darstellt. Im folgenden Algorithmus wird im Gegensatz zu HANSEN das einfache Halbierungsverfahren mit dem entwickelten Verfahren (5) gekoppelt, wobei ebenfalls Funktionswertvergleiche durchgeführt werden.

Maschinen-Intervallalgorithmus

Schritt 0: Man erstelle eine Liste (LI) mit Intervallen X_i, $X_i \in I(R_M)$, $i \in I = \{1, \ldots, m\}$, die alle Minimalpunkte der Aufgabe (10) enthalten. Dabei muß jedes Intervall nicht notwendig je ein $\hat{x}$ enthalten. Gehe zu Schritt 1.

Schritt 1: Man erstelle eine Liste (FLI) mit Intervallen $\tilde{F}_L(X_i)$, $i \in I$. Aus den Listen (LI) und (FLI) werden die Intervalle X_j und $\tilde{F}'_L(X_j)$ gestrichen, für die gilt:

$$\exists k \in I: \quad \sup \tilde{F}_L(X_k) < \inf \tilde{F}_L(X_j) \wedge j \in I \,. \tag{*}$$

Gehe zu Schritt 2.

Schritt 2: Gilt $d(X_i) \leqq \varepsilon$, $\varepsilon > 0$, für alle $i \in I$, dann werden in die Listen (LI) und (FLI), falls nicht schon vorhanden, die Zahlen $\underline{a}$, $\bar{a}$ und die Intervalle $\tilde{F}_L(\underline{a})$, $\tilde{F}_L(\bar{a})$ aufgenommen. Die Intervalle in (LI) und (FLI) werden gestrichen, für die (*) gilt. Man berechne

$$\underline{z} = \min_{1 \leqq i \leqq m} \left(\inf \tilde{F}_L(X_i)\right),$$

$$\bar{z} = \min_{1 \leqq i \leqq m} \left(\sup \tilde{F}_L(X_i)\right), \qquad \tilde{F}_L(X_i) \in \text{(FLI)}\,.$$

Man drucke die Liste (LI), deren Elemente (Intervalle) alle Minimalpunkte enthalten. Man drucke das Intervall $Z = [\underline{z}, \bar{z}]$, das $\hat{z}$ enthält.
STOP
Existiert ein X_l, $l \in I$, mit $d(X_l) > \varepsilon$, dann gehe zu Schritt 3.

Schritt 3: Man berechne $\tilde{F}'_L(X_l)$.
Ist $0 \notin \tilde{F}'_L(X_l)$, so gehe zu Schritt 5.
Ist $0 \in \tilde{F}'_L(X_l)$, so gehe zu Schritt 4.

Schritt 4: Man berechne $\tilde{F}''_L(X_l)$.
Ist $\sup \tilde{F}''_L(X_l) < 0$, so gehe zu Schritt 5.
Ist $\inf \tilde{F}''_L(X_l) > 0$, so führe das modifizierte Intervall-NEWTON-Verfahren (5') durch, wobei $X^0 := X_l$ gesetzt wird. Existiert dabei ein k_0 mit $Y^{k_0+1} \cap X^{k_0} = \emptyset$, so gehe zu Schritt 5. Das Verfahren (5') breche ab mit dem Intervall X^*, $d(X^*) \leqq \varepsilon_1 \leqq \varepsilon$. Setze $X_l := X^*$ und gehe zu Schritt 1.
Ist $0 \in \tilde{F}''_L(X_l)$, so „halbiere" X_l zu $X_{l1} = [\underline{x}_l, \tilde{m}(X_l)]$ und $X_{l2} = [\tilde{m}(X_l), \bar{x}_l]$. Anstelle X_l werden X_{l1} und X_{l2} in die Liste (LI) aufgenommen; gehe zu Schritt 1.

Schritt 5: Streiche X_l aus (LI) und $\tilde{F}_L(X_l)$ aus (FLI) und gehe zu Schritt 2.

Bemerkung 1: Der Algorithmus liefert eine Anzahl „optimalstellenverdächtiger" Intervalle. Ist die Anzahl der Minimalpunkte k, dann erzeugt der Algorithmus eine Intervallfolge, die genau k Teilfolgen enthält, die je gegen einen Optimalpunkt konvergieren. Unter Konvergenz soll in diesem Falle numerische Konvergenz verstanden werden, d. h. mit $L \to \infty$ streben die Folgen gegen die Minimalpunkte. Die Vorgabegrößen ε_1 und ε sind dabei Funktionen von L mit

$$\lim_{L \to \infty} \varepsilon_1(L) = \lim_{L \to \infty} \varepsilon(L) = 0\,.$$

Bemerkung 2: Der numerische Aufwand für ein konkretes Beispiel läßt sich a priori kaum abschätzen. Es kann z. B. sogar vorkommen, daß der Aufwand sich verringert, wenn das Restriktionsintervall vergrößert wird. Der numerische Aufwand wird dann i. a. gering sein, wenn die Anzahl der Nullstellen von $f'(x)$ klein ist und wenn $|f'(x)|$ über einem Teilintervall nicht „sehr klein" oder „sehr groß" ist.

6. Mehrdimensionaler Fall

Es wird die parametrische Aufgabe

$$\hat{z}(c) = f(\hat{x}(c); c) = \min f(x; c)\,, \qquad c \in [c]\,, \quad \text{bei} \quad x \in [x] \tag{12}$$

betrachtet, wobei $f: D \times [c] \to R^1$, $[x] \subset D \subset R^n$ für $c \in [c]$ eine streng konvexe Funktion in x über $[x]$ sei. Weiterhin besitze f stetige partielle Ableitungen bis 2. Ordnung nach x. Völlig analog zur Aufgabe (7) werden die Lösungsmengen $\hat{X}$ und $\hat{Z}$ von (12) definiert. Zur übersichtlicheren Schreibweise wird im folgenden die Abkürzung

$$FU_i^k := F'_{x_i}(X_1^{k+1}, \ldots, X_{i-1}^{k+1}, m(X_i^k), X_{i+1}^k, \ldots, X_n^k; [c])$$

benutzt. Die Aufgabe (12) wurde nur in Spezialfällen (man vgl. OELSCHLÄGEL und SÜSSE [11]) in der Literatur bisher betrachtet.

Satz 7: *Unter der Voraussetzung*

$$\inf F''_{x_i x_i}([x]; [c]) > 0 \quad \textit{für} \quad i = 1, 2, \ldots, n$$

erzeugt mit $[x]^0 := [x]$ *das Iterationsverfahren*

$$\left.\begin{array}{l}
U_i^k = \begin{cases} [\underline{x}_i^k,\ m(X_i^k)], & \textit{falls}\ \ \inf FU_i^k > 0 \\ X_i^k \quad , & \textit{falls}\ \ 0 \in FU_i^k \quad , \\ [m(X_i^k), \bar{x}_i^k], & \textit{falls}\ \ \sup FU_i^k < 0 \end{cases} \\
M_i^k = F''_{x_i x_i}(X_1^{k+1}, \dots, X_{i-1}^{k+1}, U_i^k, X_{i+1}^k, \dots, X_n^k; [c]) \\
Y_i^{k+1} = m(X_i^k) - \dfrac{FU_i^k}{M_i^k}, \qquad Z_i^{k+1} = Y_i^{k+1} \cap U_i^k , \\
X_i^{k+1} = \begin{cases} Z_i^{k+1}, & \textit{falls}\ \ Z_i^{k+1} \neq \emptyset \\ \underline{x}_i^k, & \textit{falls}\ \ \bar{y}_i^{k+1} < \underline{u}_i^k \\ \bar{x}_i^k, & \textit{falls}\ \ \underline{y}_i^{k+1} > \bar{u}_i^k \end{cases}
\end{array}\right\}, \quad i = 1(1)\,n, \quad k = 0, 1, 2, \dots, \tag{13}$$

eine Folge von Intervallvektoren $\{[x]^k\}$. *Abgebrochen wird die Iteration, falls* $Z_i^{k_0+1} = \emptyset$ *für* $i = 1(1)n$ *ist. Für die nach* (13) *erzeugte Folge* $\{[x]^k\}$ *gilt:*

(B1) $\hat{X} \subseteq [x]^k, k \geqq 0$;

(B2) $\{[x]^k\}$ *ist monoton;*

(B3) *ist* $\underline{y}_i^{k_1+1} > \bar{x}_i^{k_1}$, *so folgt* $\hat{x}_i = \bar{x}_i^{k_1}$ *für alle* $c \in [c]$,
ist $\bar{y}_i^{k_2+1} < \underline{x}_i^{k_2}$, *so folgt* $\hat{x}_i = \underline{x}_i^{k_2}$ *für alle* $c \in [c]$;

(B4) $\lim\limits_{k\to\infty} [x]^k = [x]^*$ *mit* $[x]^* \supseteq \hat{X}$, *wobei* $[x]^*$ *i.a. nur Oberhülle von* $\hat{X}$ *ist.*

Beweis:

Zu (B2): Die Monotonie $[x]^0 \supseteq [x]^1 \dots$ ist leicht einzusehen.

Zu (B1): Es wird ein beliebiges Element $x^k \in [x]^k$ und ein beliebiges Element $c \in [c]$ betrachtet, wobei man sich in (13) $[x]^k$ durch x^k ersetzt denke. Dann geht (13) für ein $i \in \{1, 2, \dots, n\}$ in die Iteration (5) über, wobei aus Satz 1 folgt, daß $\tilde{x}_i^k \in X_i^{k+1}$ gilt. Dabei ist $\tilde{x}_i^k = \min\limits_{x_i \in X_i^k} f(x_1^k, \dots, x_{i-1}^k, x_i, x_{i+1}^k, \dots, x_n^k; c)$. Wird speziell $x^k = \hat{x} \in [x]^k$ gesetzt, so folgt daraus $\hat{x}_i \in X_i^{k+1}$ und somit für $i = 1, 2, \dots, n$ die Relation $\hat{x} \in [x]^{k+1}$. Durch vollständige Induktion folgt dann $\hat{x}(c) \in [x]^k$ für $k = 0, 1, 2, \dots$

Zu (B3): Folgt sofort aus Satz 1.

Zu (B4): Folgt aus (B1) und (B2).

Bemerkung: Das Newton-Verfahren für Gleichungssysteme ist nicht unmittelbar analog (13) modifizierbar. Ist aber z. B. $N(x, [x]; [c]) \subseteq [x]$ erfüllt, dann sind nach Nickel [10] alle Minimalpunkte innere Punkte, und das Newton-Verfahren für Gleichungssysteme kann zur Einschließung der Menge $\hat{X}$ benutzt werden. $N(x, [x]; [c])$ ist dabei der Newton-Operator bezüglich $\nabla_x f(x; c)$ mit $x \in [x]$.

Ist die Aufgabe (12) nicht parametrisch, so kann folgender Konvergenzsatz angegeben werden.

Satz 8: *Zusätzlich zu den Voraussetzungen von Satz* 7 *gelte:*

$$d(F'_{x_i}([y])) \leqq L_i\, d([y]), \qquad L_i \geqq 0, \quad \textit{für alle}\ \ [y] \subseteq [x], \qquad i = 1, 2, \dots, n. \tag{14}$$

Ist dann $q = \max \dfrac{L_i}{\underline{m}_i^0} < 1$, *so folgt* $\lim\limits_{k\to\infty} [x]^k = \hat{x}$.

Der Beweis wird mit Hilfe einfacher Rechenregeln geführt, die man bei Alefeld und Herzberger [1] nachlesen kann.

7. Numerische Beispiele

Die Beispiele wurden am ORZ der TH Leuna-Merseburg auf der EDVA R 21 gerechnet. Dabei wurde ein FORTRAN-Precompiler sowie eine Mantissenlänge von 16 Dezimalen benutzt.

Beispiel 1 (Beispiel zu Aufgabe (7)):

$$f(x; c) = c_1(x - c_3)^2\, e^{c_2(x - c_3)} + (x - c_4)^2 = \min!$$

mit $c_1 \in [0.95, 1.05]$, $c_2 \in [-1.05, -0.95]$, $c_3 \in [-0.21, -0.19]$, $c_4 \in [-1.05, -0.95]$ bei $A = [-5, 0]$. Der Algorithmus (8) ergab nach 28 Iterationen

$$\hat{X} \subseteq [-0.415\dots, -0.267\dots] \quad \text{und} \quad \hat{Z} \subseteq [0.288\dots, 0.680\dots].$$

Die Voraussetzungen der Sätze 3 und 4 sind nicht erfüllt, so daß zur Verbesserung der Oberhülle Satz 5 benutzt werden soll. Dazu werden als Startintervalle nach (9) gewählt:

$$X_1^0 = [-5, -0.38], \qquad X_2^0 = [-5, -0.36], \qquad X_3^0 = [-5, -0.34];$$

$$X_1^{0'} = [-0.27, 0], \qquad X_2^{0'} = [-0.30, 0], \qquad X_3^{0'} = [-0.33, 0].$$

Es ergab sich

$$\underline{x}^{**} = -0.37864\dots, \qquad \bar{x}^{**} = -0.29859\dots.$$

Somit ist das Intervall

$$[-0.37864\ldots, -0.29859\ldots]$$

eine wesentlich günstigere Oberhülle von $\hat{X}$.

Beispiel 2 (Beispiel zu Aufgabe (10)):

$$f(x) = x^4 - 12x^3 + 41x^2 - 18(x+4) + 10(x-8)^{-5} = \min!$$

Dieses Beispiel wurde aus Jacobsen und Torabi [8] entnommen.

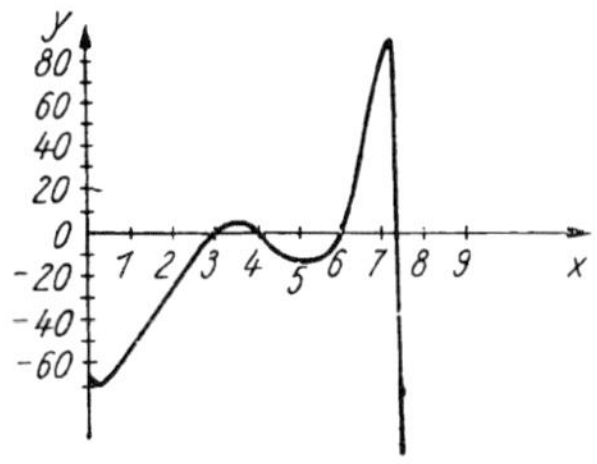

Abb. 1

Anwendung des Algorithmus aus Abschnitt 5:

a) bei $A = [0, 7.5]$.

Schon bei $\varepsilon = 0.1$ wurde das globale Minimum zu

$$\hat{x} \in [7.499999999999999, 7.500000000000001], \quad \hat{z} \in [-119.1875000001426, -119.1874999999716]$$

ermittelt. Das Verfahren (5) wurde dabei 8-mal aufgerufen, wobei 6-mal die Durchschnittsbildung auf die leere Menge führte.

b) bei $A = [0, 7]$.

Mit $\varepsilon = 0.1$ ergab sich:

$$\hat{x} \in [0.2451890984542211, 0.2451890984542242], \quad \hat{z} \in [-74.12220318862364, -74.12220318862215].$$

Das Verfahren (5) wurde dabei 9-mal aufgerufen, wobei 7-mal die Durchschnittsbildung auf die leere Menge führte.

Beispiel 3 (Beispiel zu Aufgabe (12)):

$$f(x) = \sum_{i=1}^{n} x_i^2 + \sum_{i=1}^{n-1} x_i^2 x_{i+1}^4 = \min!$$

bei $x \in [x]$, wobei $X_i = [-2, 2.10 + 0.1i]$, $i = 1(1)\, n$ ist.

Die exakten Lösungen sind stets $\hat{x}^T = (0, 0, \ldots, 0)$ und $z = 0$. Dieses Beispiel wurde aus Dussel [3], Seite 6 entnommen. Dieses Beispiel wurde dort mit einem Halbierungsalgorithmus gelöst, wobei $n = 2(1)10$ ist und die EDVA EL X8 der Universität Karlsruhe benutzt wurde. Die Rechenzeiten dieser Anlage sind absolut mit denen der R 21 nicht vergleichbar. Vergleichbar ist aber die Abhängigkeit der Rechenzeit von der Dimension der Aufgabe, deshalb wurde für $n = 2$ eine Normierung der Rechenzeiten zu einer Einheit vorgenommen. Die Komponenten der optimalen Lösung wurden dabei mit einer Genauigkeit $< 10^{-15}$ berechnet.

Tabelle 1

Algorithmus von		
Dimension n	Dussel EDVA EL X8	Algorithmus (13) EDVA R 21
2	1	1
3	2.1	1.16
4	3.5	1.23
5	5.9	1.42
6	9.4	1.55
7	19.6	1.70
8	37.1	1.81
9	87.1	1.96
10	210.8	2.07

8. Literatur

1 Alefeld, G.; Herzberger, J., Einführung in die Intervallrechnung, Bibl. Mannheim 1974.

2 Brossart, F. J., Einschließung des Minimalintervalls einer konvexen Funktion auf einem Intervall aus $I(R)$, Dipl.-Arbeit, Karlsruhe 1977.

3 Dussel, R., Einschließung des Minimalpunktes einer streng konvexen Funktion auf einem n-dimensionalen Quader, Dissertation, Karlsruhe 1972.

4 HANSEN, E., A globally convergent interval method for computing and bounding real roots, BIT **18**, 415–424 (1978).
5 HANSEN, E., Global optimization using interval analysis: The one-dimensional case, J. Optimization Theory Appl. **29**, No. 3, 351–344 (1979).
6 HANSEN, E., Global optimization using interval analysis: The multidimensional case, Numer. Math. **34**, 247–270 (1980).
7 HERZBERGER, J., Bemerkungen zu einem Verfahren von MOORE, ZAMM **53**, 356–357 (1973).
8 JACOBSEN, E. S.; TORABI, M., A global minimization algorithm for a class of one-dimensional functions, J. Math. Analysis Appl. **62**, 310–324 (1978).
9 MANCINI, L. J.; WILDE, D. J., Interval arithmetic in unidimensional signomial programming, J. Optimization Theory Appl. **26**, No. 2, 277–289 (1978).
10 NICKEL, K., On the NEWTON method in interval analysis, Technical Summary Report No. 1136, Mathematics Research Center, University of Wisconsin, Madison, Wisc. 1971.
11 OELSCHLÄGEL, D.; SÜSSE, H., Interval analytic treatment of convex programming problems, Computing **24**, 213–225 (1980).
12 OELSCHLÄGEL, D.; SÜSSE, H., Numerische Untersuchungen zu einigen Intervall-NEWTON-Verfahren, (To appear in: Wiss. Z. TH Leuna-Merseburg).

Eingereicht am 3. 3. 1980, revidierte Fassung am 18. 8. 1980

Anschrift: Prof. Dr. DIETER OELSCHLÄGEL, Dr. HERBERT SÜSSE, Technische Hochschule „Carl Schorlemmer" Leuna-Merseburg, Sektion Mathematik, DDR-4200 Merseburg, Geusaer Straße

ZAMM 61, 253 – 260 (1981)

K. STREHMEL

Stabilitätseigenschaften adaptiver Runge-Kutta-Verfahren

Zur numerischen Behandlung von steifen Anfangswertaufgaben wird eine Klasse adaptiver Runge-Kutta-Verfahren betrachtet. Unter Beibehaltung der Konsistenzordnung erfordern sie nur gelegentlich eine Berechnung der Jacobi-Matrix. Das Stabilitätsverhalten der adaptiven Verfahren bei großen Schrittweiten wird bezüglich des von Prothero und Robinson vorgeschlagenen skalaren Anfangswertproblems diskutiert. Die Wirksamkeit der adaptiven Verfahren wird durch einige steife Anfangswertaufgaben belegt.

A class of adapted Runge-Kutta methods is considered for the numerical solution of stiff systems of ordinary differential equations. These require only occasional calculation of the Jacobian matrix while maintaining the order of consistency. The stability behaviour of the adapted methods with great steplength will be discussed with reference to the scalar initial value problem of Prothero and Robinson. The efficiency of the adapted methods is demonstrated by several stiff-initial value problems.

Для численной реализации жестких задач Коши рассматривается один класс адаптивных методов Рунге-Кутта. Эти методи позволяют при сохранении порядка локальной ошибки дискретизации сохранять матрицу Якоби несколько шагов. Поведение устойчивости адаптивных методов при больших величинах шага обсуждается относительно одномерной задачи Коши, предложенной Просеро и Робинсоном. Эффективность адаптивных методов иллюстрируется на нескольких задачах Коши для жестких дифференциальных уравнений.

1. Einleitung

Betrachtet wird ein Anfangswertproblem für ein System steifer gewöhnlicher Differentialgleichungen erster Ordnung

$$\begin{aligned} y' &= f(x, y) \\ y(x_0) &= y_0 \end{aligned} \quad \text{mit} \quad x \in [x_0, x_E] \tag{1.1}$$

in einem q-dimensionalen reellen Vektorraum. Die eindeutige Lösbarkeit der Anfangswertaufgabe (1.1) wird dabei vorausgesetzt.

Der vorliegende Beitrag befaßt sich mit der Untersuchung der Stabilität bei fester Schrittweite von s-stufigen adaptiven RUNGE-KUTTA-Verfahren. Es werden notwendige und hinreichende Bedingungen angegeben, um S-stabile und L-S-stabile adaptive RUNGE-KUTTA-Verfahren zu erhalten. Eine Klasse zweistufiger intern S-stabiler Verfahren sowie ein zweistufiges und ein vierstufiges intern S- und L-S-stabiles Verfahren werden mitgeteilt.

Die vorliegende Klasse von Einschrittverfahren beruht bekanntlich auf einer geeigneten Linearisierung des vorgelegten Differentialgleichungssystems in jedem Integrationsschritt (vgl. FRIEDLI [4]). Das lineare System mit einer konstanten (q, q)-Matrix T wird dann analytisch gelöst. Wir untersuchen demzufolge auch das Stabilitätsverhalten bei fester Schrittweite, wenn bei nichtlinearen Systemen die Matrix T des linearisierten Differentialgleichungssystems nur von „Zeit zu Zeit" der momentanen JACOBI-Matrix von $f(x, y)$ angepaßt wird. Schließlich geben wir einige numerische Beispiele für steife Systeme.

Andere Modifikationen von verallgemeinerten Runge-Kutta Verfahren findet man bei LAWSON [8] sowie be- VAN DER HOUWEN [6]. Letztgenannte Verfahrensklasse wurde bezüglich S-Stabilität von VERWER [13] untersucht.

2. Adaptive Runge-Kutta-Verfahren

Ein s-stufiges adaptives RUNGE-KUTTA-Verfahren (ARK(s)-Verfahren) mit dem Parameterschema

$$\begin{array}{l|lllll} c_1 = 0 & & & & & \\ c_2 & A_{21}(c_2hT) & & & & \\ c_3 & A_{31}(c_3hT) & A_{32}(c_3hT) & & & \\ \vdots & \vdots & & & & \\ c_s & A_{s1}(c_shT) & A_{s2}(c_shT) & \dots & A_{s\,s-1}(c_shT) & \\ \hline c_{s+1} = 1 & A_{s+1,1}(hT) & A_{s+1,2}(hT) & \dots & A_{s+1,s-1}(hT) & A_{s+1,s}(hT) \end{array} \tag{2.1}$$

(vgl. FRIEDLI [4]) liefert für die wahre Lösung des Anfangswertproblems (1.1) in den Gitterpunkten $x_{m+1} = x_m + h$, $m = 0, 1, \dots$, die Näherungen

$$u_{m+1} := e_0(hT)\, u_m + h \sum_{i=1}^{s} A_{s+1,i}(hT)\, r(x_{m+1}^{(i)}, u_{m+1}^{(i)}) \tag{2.2}$$

mit

$$x_{m+1}^{(i)} := x_m + c_i h; \qquad u_{m+1}^{(i)} := e_0(c_ihT)\, u_m + h \sum_{l=1}^{i-1} A_{il}(c_ihT)\, r(x_{m+1}^{(l)}, u_{m+1}^{(l)});$$

$$r(x_{m+1}^{(i)}, u_{m+1}^{(i)}) := f(x_{m+1}^{(i)}, u_{m+1}^{(i)}) - T \cdot u_{m+1}^{(i)}, \qquad i = 1, 2, \dots, s.$$

Dabei ist T eine auf dem Intervall $[x_m, x_{m+1}]$ beliebige konstante (q, q)-Matrix und die Matrizenfunktionen $A_{ij}(c_ihT)$ sind durch

$$A_{il}(c_ihT) := \sum_{j=0}^{\varrho_i} a_{il}^{(j)} e_{j+1}(c_ihT); \qquad \varrho_i \leqq \varrho_{i+1},$$

mit

$$e_0(B) := \exp(B); \qquad B := c_i hT ,$$
$$e_1(B) := [e_0(B) - I]\, B^{-1} , \qquad I: q\text{-reihige Einheitsmatrix}$$
$$e_{j+1}(B) := [j \cdot e_j(B) - I]\, B^{-1} , \qquad j = 1, 2, \dots ,$$

definiert. Für $T = 0$ reduziert sich (2.1) auf ein s-stufiges explizites RUNGE-KUTTA-Verfahren (RK(s)-Verfahren).

Bemerkung 1: Ein ARK(s)-Verfahren beruht im Prinzip auf der formalen Lösung des Differentialgleichungssystems

$$y' = Ty + r(x, y) \tag{2.3}$$

indem in jedem Diskretisierungsintervall $[x_m, x_{m+1}^{(i)}]$, $i = 2, 3, \dots, s + 1$; die Funktion $r(x, y)$ durch ein Vektorpolynom

$$r_h^{(i)}(x) = \sum_{n=0}^{\varrho_i} b_{n,m}^{(i)}(x - x_m)^n; \tag{2.4}$$

approximiert wird. Die Koeffizienten $b_{n,m}^{(i)}$ sind dabei von den Funktionswerten $r_l := r(x_{m+1}^{(l)}, u_{m+1}^{(l)})$, $l = 1, 2, \dots, i - 1$, und der Schrittweite h abhängig.

Die Lösungsdarstellung (2.2) ist jedoch für eine numerische Rechnung kaum geeignet, da sie die Bereitstellung der Exponentialmatrix $e_0(B)$ bzw. die zugehörige Matrixreihe — die an geeigneter Stelle abzubrechen ist — erfordert. Außerdem sind die Funktionen $e_j(B)$, $j = 1, 2, \dots$, für $\det(B) = 0$ „numerisch singulär" (ein rationaler Ausdruck mit einer hebbaren Singularität heißt bei hinreichend kleinem Nenner ,,numerisch singulär"), so daß ihre Berechnung für $\det(B) \to 0$ zu numerischen Schwierigkeiten führt. Daher verwenden wir im folgenden für die Funktionen $e_j(z)$, $z \in \mathbb{C}$, $j = 0, 1, \dots$, die Approximationen

$$e_0(z) \approx R_0(z) , \qquad e_1(z) \approx [R_0(z) - 1]/z =: R_1(z) , \tag{2.5}$$
$$e_{j+1}(z) \approx [jR_j(z) - 1]/z =: R_{j+1}(z) , \qquad j = 1, 2, \dots ,$$

wobei $R_0(z)$ eine rationale Funktion ist. *Besitzt $R_0(z)$ die Approximationsordnung $k \geqq \varrho_{s+1}$, so haben die Funktionen $R_j(z)$, $j = 0, 1, \dots, k + 1$, den gleichen Nenner* (der Beweis ist trivial). *Ein* ARK(s)-*Verfahren ist* demzufolge *explizit, wenn $R_0(z)$ ein Polynom ist, sonst ist es semi-implizit.* Im letzten Fall sind in jedem Integrationsschritt, in dem die Matrix T geändert wird, $(\mu + 1)$-LU-Zerlegungen erforderlich, wobei μ die Anzahl der verschiedenen Schrittparameter c_i, $i = 2, 3, \dots, s$, mit $c_i \neq 1$ des ARK(s)-Verfahrens ist.

3. Konsistenz der ARK(s)-Verfahren

Für ein ARK(s)-Verfahren gilt der folgende Konsistenzsatz:

Satz 1: *Ist $f(x, y)$ p-mal stetig differenzierbar in*

$$U := \{(x, y) \mid x_0 \leqq x \leqq x_E, \|y - y(x)\| \leqq \sigma, \sigma > 0\}$$

und gilt

$$\int_{x_m}^{x_m+h} [r(t, y(t)) - r_h^{(s+1)}(t)]\, dt = O(h^{p+1}), \qquad h \to 0 ,$$

dann hat ein ARK(s)-*Verfahren* (2.1) *die Konsistenzordnung p, wenn die Stabilitätsfunktion $R_0(z)$ eine Ordnung $k \geqq \max(\varrho_{s+1}, p)$ besitzt.*

Beweis: Aus (2.3) folgt

$$y(x_{m+1}) = e_0(hT)\, y(x_m) + \int_{x_m}^{x_{m+1}} e_0(T(x_m + h - t))\, r(t, y(t))\, dt .$$

Mit (2.4) und (2.5) ergibt sich die Näherungslösung u_{m+1} in der Form

$$u_{m+1} = R_0(hT)\, u_m + h \sum_{n=0}^{\varrho_s+1} R_{n+1}(hT)\, b_{n,m}^{(s+1)} h^n .$$

Der lokale Diskretisierungsfehler $\tau(x_m; h)$ ist damit durch

$$\tau(x_m; h) = h^{-1} \{[e_0(hT) - R_0(hT)]\, y(x_m) + \int_{x_m}^{x_{m+1}} e_0(T(x_m + h - t))\, r(t, y(t))\, dt - h \sum_{n=0}^{\varrho_s+1} R_{n+1}(hT)\, b_{n,m}^{(s+1)} h^n\}$$

gegeben. Wegen

$$R_n(hT) = e_n(hT) + O[(hT)^{k+1-n}] \quad \text{für} \quad \|hT\| \to 0$$

erhält man

$$\tau(x_m; h) = h^{-1} \left\{ \int_{x_m}^{x_{m+1}} e_0(T(x_m + h - t))\, r(t, y(t))\, dt - h \sum_{n=0}^{\varrho_s+1} e_{n+1}(hT)\, b_{n,m}^{(s+1)} h^n \right\} + O(h^k) ,$$

$$= h^{-1} \int_{x_m}^{x_{m+1}} e_0(T(x_m + h - t)\, [r(t, y(t)) - r_h^{(s+1)}(t)]\, dt + O(h^k) , \qquad h \to 0 . \tag{3.1}$$

Setzt man die Entwicklung

$$e_0(T(x_m + h - t)) = I + T(x_m + h - t) + \dots$$

in (3.1) ein, so liefert der erste Summand den Beitrag

$$h^{-1} \int_{x_m}^{x_{m+1}} [r(t, y(t)) - r_h^{(s+1)}(t)] \, dt , \tag{3.2}$$

der voraussetzungsgemäß die Ordnung p hat. Die weiteren Summanden im Integralterm von (3.1) liefern Beiträge, die von höherer Ordnung in h sind, als (3.2).

4. Stabilität der ARK(s)-Verfahren bei fester Schrittweite

Da die Eigenschaft der A-Stabilität (vgl. DAHLQUIST [2]) eines numerischen Verfahrens bei der Behandlung sehr steifer Differentialgleichungssysteme mit variabler JACOBI-Matrix noch keine stabile Näherungslösung garantiert, führten PROTHERO und ROBINSON [9] den stärkeren Begriff der S-Stabilität ein.

Definition 1: Ein Diskretisierungsverfahren heißt *S-stabil*, wenn es bei Anwendung auf die skalare Testdifferentialgleichung

$$y'(x) = \lambda(y(x) - g(x)) + g'(x) , \qquad \lambda \in \mathbb{C} , \qquad \mathrm{Re}(\lambda) < 0 , \qquad g(x) \in C^1[0, x^*] \tag{4.1}$$

eine Folge von Näherungen $\{u_m\}$ mit der Eigenschaft liefert, daß zu jedem $\lambda_0 < 0$ ein $h_0 > 0$ existiert, so daß für alle Schrittweiten $h \in (0, h_0)$ und alle λ mit $\mathrm{Re}(\lambda) \leqq \lambda_0$ gilt

$$\left| \frac{u_{m+1} - g(x_{m+1})}{u_m - g(x_m)} \right| < 1 \quad \text{mit} \quad x_m, x_{m+1} \in [0, x^*] ,$$

falls $u_m \neq g(x_m)$ ist.

Wir untersuchen jetzt das Stabilitätsverhalten der ARK(s)-Verfahren bezüglich der Testgleichung (4.1). Aus

$$u_{m+1}^{(i)} = R_0(c_i h T) \, u_m + h c_i \sum_{n=0}^{\varrho_i} R_{n+1}(c_i h T) \, b_{n,m}^{(i)} (c_i h)^n , \qquad i = 2, 3, \dots, s+1 ,$$

$(u_{m+1} := u_{m+1}^{(s+1)})$ folgt bei Anwendung auf (4.1)

$$u_{m+1}^{(i)} = R_0(c_i z) \, u_m + h c_i T_m^{(i)}(z, h; g) \tag{4.2}$$

mit

$$T_m^{(i)}(z, h; g) = \sum_{n=0}^{\varrho_i} R_{n+1}(c_i z) \, b_{n,m}^{(i)} (c_i h)^n , \qquad z = h\lambda .$$

Wegen

$$b_{0,m}^{(i)} = \sum_{l=1}^{i-1} c_{0l}^{(i)} r_l \quad \text{mit} \quad \sum_{l=1}^{i-1} c_{0l}^{(i)} = 1 , \qquad c_{0l}^{(i)} \in \mathbb{R} , \tag{4.3}$$

und

$$(c_i h)^n \, b_{n,m}^{(i)} = \sum_{l=1}^{i-1} c_{n,l}^{(i)} r_l , \qquad n = 1, 2, \dots, \varrho_i , \tag{4.4}$$

mit

$$\sum_{l=1}^{i-1} c_{n,l}^{(i)} = 0 , \qquad c_{n,l}^{(i)} \in \mathbb{R} ,$$

folgt

$$T_m^{(i)}(z, h; g) = \sum_{n=0}^{\varrho_i} R_{n+1}(c_i z) \sum_{l=1}^{i-1} c_{n,l}^{(i)} \delta(x_m + c_l h) , \tag{4.5}$$

wobei zur Abkürzung

$$\delta(x) = g'(x) - \lambda g(x)$$

gesetzt wurde. Für den Diskretisierungsfehler

$$\varepsilon_{m+1}^{(i)} := u_{m+1}^{(i)} - g(x_m + c_i h) , \qquad i = 1, 2, \dots, s+1 ,$$

erhält man dann in der i-ten Stufe, $i = 2, 3, \dots, s+1$, die Differenzengleichung

$$\varepsilon_{m+1}^{(i)} = R_0(c_i z) \, \varepsilon_{m+1}^{(1)} + h c_i \left[T_m^{(i)}(z, h; g) + \frac{1}{c_i h} (R_0(c_i z) \, g(x_m) - g(x_m + c_i h)) \right] . \tag{4.6}$$

Es ergibt sich nun der

Satz 2: *Ein konsistentes* ARK(s)-*Verfahren ist genau dann S-stabil, wenn die Stabilitätsfunktion* $R_0(z)$ *der Ordnung* $k \geqq \varrho_{s+1}$ *stark A-akzeptabel ist.*

Beweis: Wir verwenden das

Kriterium (vgl. VERWER [11]): *Ein s-stufiges konsistentes Einschrittverfahren, das bei Anwendung auf* (4.1 *eine Fehlergleichung der Form*

$$\varepsilon_{m+1}^{(i)} = T_i(z) \, \varepsilon_{m+1}^{(1)} + h \left[\Psi_m^{(i)}(z, h; g) + \frac{1}{h} (T_i(z) \, g(x_m) - g(x_m + c_i h)) \right] , \qquad i = 2, 3, \dots, s+1 ,$$

liefert, ist genau dann S-stabil, wenn

a) *die rationale Stabilitätsfunktion* $T_{s+1}(z)$ *stark A-akzeptabel ist,*

b) *eine Konstante* $\bar{h} > 0$ *existiert, so daß*

$$\Psi_m^{(s+1)}(z, h; g) + \frac{1}{h}\left(T_{s+1}(z)\, g(x_m) - g(x_m + h)\right)$$

auf $M := \{(h, z) \mid \operatorname{Re}(z) < 0, h \in (0, \bar{h}]\}$ *gleichmäßig beschränkt ist.*

Die Anwendung des Mittelwertsatzes der Differentialrechnung auf

$$\xi_m^{(s+1)} := T_m^{(s+1)}(z, h; g) + \frac{1}{h}\left(R_0(z)\, g(x_m) - g(x_m + h)\right)$$

liefert mit (4.3), (4.4) und (4.5)

$$\xi_{m+1}^{(s+1)} = \sum_{n=0}^{\varrho_s+1} R_{n+1}(z) \sum_{l=1}^{s} c_{n,l}^{(s+1)} g'(x_m + c_l h) - \sum_{n=0}^{\varrho_s+1} z R_{n+1}(z) \sum_{l=2}^{s} c_{n,l}^{(s+1)} g'(x_m + \theta_l c_l h)\, c_l - g'(x_m + \theta_{s+1} h)$$

mit $0 < \theta_j < 1, j = 2, \dots, s+1$. Da die Funktionen $R_n(z)$ und $zR_n(z)$, $n = 1, 2, \dots, \varrho_{s+1} + 1$, auf $\{z \mid \operatorname{Re}(z) < 0\}$ gleichmäßig beschränkt sind, ist $\xi_m^{(s+1)}$ auf M gleichmäßig beschränkt.

Für Verfahren zur Lösung sehr steifer Differentialgleichungen ist es zweckmäßig, die Eigenschaft der S-Stabilität für jeden Teilschritt $u_{m+1}^{(i)}$, $i = 2, 3, \dots, s+1$, zu fordern. Derartige Verfahren nennt VERWER [12] *intern* S-stabil.

Wegen $c_i \neq 0$, $i = 2, 3, \dots, s$, erhält man unmittelbar den

Satz 3: *Jedes S-stabile* ARK(s)-*Verfahren ist intern S-stabil.*

Definition 2: Ein S-stabiles Diskretisierungsverfahren heißt *L-S-stabil,* wenn

$$\left|\frac{u_{m+1} - g(x_{m+1})}{u_m - g(x_m)}\right| \to 0\,, \qquad \operatorname{Re}(\lambda) \to -\infty$$

für alle Schrittweiten h mit $x_m, x_{m+1} \in [0, x^*]$ gilt.

Es ergibt sich der

Satz 4: *Ein S-stabiles* ARK(s)-*Verfahren ist genau dann L-S-stabil, wenn die Stabilitätsfunktion* $R_0(z)$ *L-akzeptabel ist und die Bedingungen*

$$\lim_{\operatorname{Re}(z) \to -\infty} z \sum_{\mu} A_{s+1,\mu}(z) = \begin{cases} 0 & \text{für } c_\mu = a\,, \quad 0 \leqq a < 1 \\ -1 & \text{für } |c_\mu = 1| \end{cases} \quad \textit{mit} \quad \mu \in \{1, 2, \dots, s\}$$

erfüllt sind.

Beweis: Nach (4.6) und Definition 2 liegt L-S-Stabilität genau dann vor, wenn für jedes $h > 0$ mit $x_m, x_{m+1} \in [0, x^*]$

$$\lim_{\operatorname{Re}(z) \to -\infty} R_0(z) = 0\,, \quad \lim_{\operatorname{Re}(z) \to -\infty} \left[T_m^{(s+1)}(z, h; g) - \frac{1}{h} g(x_m + h)\right] = 0$$

gilt. Nach (2.2) und (4.2) mit (2.5) ist

$$T_m^{(s+1)}(z, h; g) = \sum_{l=1}^{s} A_{s+1,l}(z)\, [g'(x_m + c_l h) - \lambda g(x_m + c_l h)]\,.$$

Wegen

$$\lim_{\operatorname{Re}(z) \to -\infty} A_{s+1,l}(z) = 0 \quad \text{für} \quad l = 1, 2, \dots, s$$

ergibt sich

$$\lim_{\operatorname{Re}(z) \to -\infty} \left[T_m^{(s+1)}(z, h; g) - \frac{1}{h} g(x_m + h)\right] = -\frac{1}{h}\left[\lim_{\operatorname{Re}(z) \to -\infty} \sum_{l=1}^{s} z A_{s+1,l}\, g(x_m + c_l h) + g(x_m + h)\right],$$

woraus unmittelbar die Behauptung folgt.

Bemerkung 2:

1. Das ARK(2)-Verfahren (vgl. FRIEDLI [4])

$$\begin{array}{c|cc} 0 & & \\ c_2 & c_2 R_1(c_2 hT) & \\ \hline & R_1(hT) - \dfrac{1}{c_2} R_2(hT) & \dfrac{1}{c_2} R_2(hT) \end{array} \tag{4.7}$$

ist, wenn $R_0(z)$ der Ordnung $k \geqq 1$ L-akzeptabel ist, für $c_2 = 1$ intern S- und L-S-stabil. Für $0 < c_2 < 1$ ist das Verfahren intern S-stabil.

2. Das ARK(4)-Verfahren, das auf dem klassischen RK(4)-Verfahren basiert,

$$\begin{array}{l|llll}
0 & & & & \\
1/2 & \frac{1}{2} R_1(\frac{1}{2} hT) & & & \\
1/2 & \frac{1}{2} R_1(\frac{1}{2} hT) - R_2(\frac{1}{2} hT) & R_2(\frac{1}{2} hT) & & \\
1 & R_1(hT) - 2R_2(hT) & 0 & 2R_2(hT) & \\
\hline
 & A_{51}(hT) & A_{52}(hT) & A_{53}(hT) & A_{54}(hT)
\end{array}$$

mit

$$A_{51}(hT) = R_1(hT) - 3R_2(hT) + 2R_3(hT)\,, \qquad A_{52}(hT) = 2R_2(hT) - 2R_3(hT)\,,$$

$$A_{53}(hT) = 2R_2(hT) - 2R_3(hT)\,, \qquad A_{54}(hT) = -R_2(hT) + 2R_3(hT)$$

ist — wenn $R_0(z)$ mit $k \geqq 3$ L-akzeptabel ist — intern S- und L-S-stabil. Es besitzt, wie man mit dem Konsistenzsatz leicht nachweist, für $k \geqq 4$ die Konsistenzordnung $p = 4$.

Im weiteren wird das Stabilitätsverhalten der ARK(s)-Verfahren bezüglich des linearen Testsystems

$$y' = Jy \tag{4.8}$$

untersucht, wobei die Eigenwerte der konstanten (q, q)-Matrix J einen negativen Realteil haben. Bei Anwendung von (2.1) auf (4.8) folgt

$$u_{m+1} = R_0(hJ)\, u_m\,,$$

wenn $T = J$ ist. Für nichtlineare Systeme wird man jedoch, um den Rechenaufwand zu reduzieren, die JACOBI-Matrix nicht in jedem Integrationsschritt neu berechnen, so daß das Stabilitätsverhalten der ARK(s)-Verfahren nur approximativ durch die Funktion R_0 bestimmt ist. Wir betrachten daher den Fall

$$T \approx J\,,$$

wobei das Spektrum $\Lambda(T)$ ebenfalls in der negativen komplexen Halbebene liegen soll. Bei Anwendung von (2.1) auf (4.8) ergibt sich dann

$$u_{m+1}^{(i)} = R_0(c_i hT)\, u_m + h \sum_{l=1}^{i-1} A_{il}(c_i hT)\, (J - T)\, u_{m+1}^{(l)}\,, \qquad i = 2, 3, \dots, s+1\,. \tag{4.9}$$

Um eine Schranke für die Lösung $u_{m+1}^{(s+1)}$ von (4.9) herzuleiten, benötigen wir das folgende

Lemma: *Genügt die Folge von Vektoren* $\{y_m\}$ *der homogenen linearen Differenzengleichung*

$$y_{m+1} = (C + h \cdot D)\, y_m\,, \qquad m = 0, 1, \dots,$$

wobei die Matrizen C *und* D *die Bedingungen*

$$\|C^l\|_\infty \leqq K \cdot e^{-\alpha l}\,, \qquad l = 0, 1, \dots, \qquad \|D\|_\infty \leqq L\,, \tag{4.10}$$

mit den Konstanten α, $L > 0$ *und* $K \geqq 1$ *erfüllen, dann gilt*

$$\|y_{m+1}\|_\infty \leqq K \cdot \exp\big((m+1)\,(KhL\, e^\alpha - \alpha)\big) \qquad \|y_0\|_\infty\,. \tag{4.11}$$

Das Lemma ist eine Spezifizierung eines Lemmas von STRANG [10].

Bemerkung 3: Ist der Spektralradius $\sigma(C) < 1$, dann ist die Konstante α in (4.10) positiv, und aus (4.11) folgt

$$\lim_{m \to \infty} \|y_{m+1}\|_\infty = 0 \quad \text{für} \quad h \cdot L < \frac{\alpha}{K} e^{-\alpha}\,.$$

Es ergibt sich der

Satz 5: *Die Stabilitätsfunktion* $R_0(z)$ *eines konsistenten* ARK(s)-*Verfahrens sei* A-*akzeptabel und besitze die Ordnung* $k \geqq \varrho_{s+1}$. *Dann existieren positive Konstanten* $K \geqq 1$, $\alpha > 0$, *und zu jedem vorgegebenen* $\bar{h} > 0$ *existiert ein* $\gamma_0 > 0$, *so daß mit* $d = \|J - T\|_\infty$ *für die Lösung* u_{m+1} *des* ARK(s)-*Verfahrens die Abschätzung*

$$\|u_{m+1}\|_\infty \leqq K \cdot \exp\big((m+1)\,(Khd\gamma_0\, e^\alpha - \alpha)\big)\, \|u_0\|_\infty\,, \qquad h \in (0, \bar{h}]\,, \qquad m = 0, 1, \dots, \tag{4.12}$$

gilt.

Beweis: Aus (4.9) folgt für u_{m+1} die Darstellung

$$u_{m+1} = [R_0(hT) + h \cdot Q]\, u_m\,,$$

wobei die Matrix

$$Q = Q\big(R_0(c_\nu hT)\,,\ A_{il}(c_i hT),\, J - T,\, h\big)$$

für $\nu = 2, 3, \dots, s$; $i = 2, 3, \dots, s+1$; $l = 1, 2, \dots, i-1$; die Eigenschaften

$$\|Q(R_0(c_\nu hT), A_{il}(c_i hT), 0, h)\|_\infty = 0\,, \qquad \|Q(R_0(0), A_{il}(0), J - T, 0)\|_\infty > 0 \tag{4.13}$$

besitzt. Da $R_0(z)$ A-akzeptabel ist und die Approximationsordnung $k \geqq \varrho_{s+1}$ hat, existieren positive Konstanten γ_1 und δ_{il}, $i = 2, 3, \dots, s+1$, $l = 1, 2, \dots, i-1$, so daß für jedes $h > 0$ die Abschätzungen

$$\|R_0(c_\nu hT)\|_\infty \leqq \gamma_1\,, \qquad \|A_{il}(c_i hT)\|_\infty \leqq \delta_{il}$$

gelten. Auf Grund von (4.13) gilt

$$||Q||_\infty \leqq d \cdot \gamma_0 \quad \text{für} \quad h \in (0, \bar{h}] .$$

Weiterhin ist $\sigma(R_0(hT)) < 1$, so daß für jedes $h > 0$

$$||R_0(hT)^m||_\infty \leqq K \cdot e^{-\alpha m} \quad \text{mit} \quad K \geqq 1 , \quad \alpha > 0 , \quad m = 0, 1, \ldots ,$$

gilt. Die Anwendung des Lemmas ergibt die Abschätzung (4.12).

Folgerung: *Aus* (4.12) *folgt*

$$\lim_{m\to\infty} u_{m+1} = 0 \quad Khd\gamma_0\, e^{\alpha} - \alpha < 0$$

Ein ARK(s)-Verfahren zeigt folglich für $T \approx f_y(x_m, u_m)$ ein zuverlässiges Stabilitätsverhalten, so daß es auch zur Lösung stark nichtlinearer Systeme geeignet ist.

5. Numerische Beispiele

Für die numerische Integration zweier Beispiele für steife Systeme verwenden wir das ARK(2)-Verfahren (4.7) mit dem Parameter $c_2 = \frac{2}{3}$. Als Stabilitätsfunktion $R_0(z)$ wird die L-verträgliche (2,1)-Padé-Approximation von exp (z)

$$R_0(z) := \frac{1 + \frac{1}{3} z}{1 - \frac{2}{3} z + \frac{1}{6} z^2} \tag{5.1}$$

gewählt. Das ARK(2)-Verfahren hat für beliebige Matrizen T die Konsistenzordnung $p = 2$, und für $T = f_y(x_m, u_m)$ ist $p = 3$. Es ist semi-implizit und gemäß Satz 2 und 3 intern S-stabil. Verglichen werden die numerischen Ergebnisse mit denen, die sich nach dem 4-stufigen Lawson-Verfahren „LAW(4)" mit der Stabilitätsfunktion (5.1) — bei dem das klassische RK(4)-Verfahren der Ordnung $p = 4$ verwendet wurde (vgl. Grigorieff [5]) — und dem zweistufigen Rosenbrock-Verfahren „ROS(3)" (vgl. Albrecht [1]) ergeben. Beide Verfahren haben die Konsistenzordnung $p = 3$. Das LAW(4)-Verfahren ist nicht S- sondern lediglich L-stabil, das ROS(3)-Verfahren ist intern S-stabil und besitzt die stark A-verträgliche Stabilitätsfunktion

$$R_0(z) := \frac{1 - z - \frac{2}{3} z^2}{1 - 2z + \frac{5}{6} z^2} .$$

Beide Beispiele sind nach dem gleichen Schema aufgebaut. Das Anfangswertproblem wird, beginnend in $x_0 := 0$ bis zur festen Stelle $x_E := 10$ mit den Schrittweiten $h := 2^{-k}$, $k = 0, 1, \ldots$, integriert. Sämtliche Rechnungen wurden an einer EDVA R 300 mit einer Mantissenlänge von 8 Dezimalstellen durchgeführt.

Beispiel 1 (vgl. Lapidus und Seinfeld [7]):

$$y' = g'(x) - 200(y(x) - g(x)) , \quad y(0) = 10 , \quad \text{mit} \quad g(x) = 10 - (10 + x) \exp(-x) .$$

Exakte Lösung: $y(x) = g(x) + 10 \cdot \exp(-200x)$

$$y(10) = .999\,909\,20 + 01 .$$

Numerische Ergebnisse an der Stelle $x_E := 10$:

h	ARK(2)	LAW(4)	ROS(3)
1	.99992938 +01	.30231311 +03	.11035387 +02
1/2	.99991321 +01	.13781985 +03	.10060705 +02
1/4	.99991007 +01	.58342893 +02	.99998353 +01
1/8	.99990940 +01	.22861067 +02	.99994265 +01
1/16	.99990924 +01	.10313030 +02	.99992570 +01
1/32	.99990922 +01	.83099726 +01	.99991737 +01
1/64	.99990920 +01	.91573658 +01	.99991319 +01

Verwendet man bei dem ARK(2)-Verfahren die Stabilitätsfunktion des ROS(3)-Verfahrens, so ergeben sich an der Stelle $x_E := 10$ die folgenden Näherungswerte:

h	ARK(2)
1	.10821836 +02
1/2	.10050900 +02
1/4	.99991541 +01
1/8	.99990940 +01
1/16	.99990923 +01
1/32	.99990922 +01
1/64	.99990920 +01

Beispiel 2 (vgl. ENRIGHT, HULL, LINDBERG [3]):

$$
\begin{aligned}
y_1' &= -y_1 + 2; & y_1(0) &= 1\,, \\
y_2' &= -10y_2 + \beta y_1^2; & y_2(0) &= 1\,, \\
y_3' &= -40y_3 + 4\beta(y_1^2 + y_2^2); & y_3(0) &= 1\,, \\
y_4' &= -100y_4 + 10\beta(y_1^2 + y_2^2 + y_3^2); & y_4(0) &= 1\,,
\end{aligned}
$$

1. $\beta = 1$

Exakte Lösung:

$$y_1(10) = .19999546 + 01\,, \quad y_2(10) = .39997982 + 00\,,$$
$$y_3(10) = .41597972 + 00\,, \quad y_4(10) = .43328392 + 00\,.$$

Numerische Ergebnisse an der Stelle $x_E = 10$:

h	ARK(2)	LAW(4)	ROS(3)
1	.19999596 +01	.19840871 +01	.19999741 +01
	.39998206 +00	.32913879 +00	.40178756 +00
	.41598200 +00	.82404887 +00	.45218708 +00
	.43328634 +00	−.18085148 +01	.49725841 +00
1/2	.19999553 +01	.19973459 +01	.19999585 +01
	.39998011 +00	.43074806 +00	.39998155 +00
	.41598002 +00	.91566070 −01	.41617906 +00
	.43328421 +00	−.90865530 +00	.43532360 +00
1/4	.19999547 +01	.19995787 +01	.19999552 +01
	.39997985 +00	.41633213 +00	.39998008 +00
	.41597974 +00	.46063467 +00	.41597997 +00
	.43328395 +00	−.10807983 +00	.43328433 +00

2. $\beta = 10$

Exakte Lösung:

$$y_1(10) = .19999546 + 01\,, \quad y_2(10) = .39997982 + 01\,,$$
$$y_3(10) = .19998158 + 02\,, \quad y_4(10) = .41992377 + 03\,.$$

Numerische Ergebnisse an der Stelle $x_E = 10$:

h	ARK(2)	LAW(4)	ROS(3)
1	.19999596 +01	.19840841 +01	.19999741 +01
	.39998206 +01	.32913597 +01	.40010503 +01
	.19998361 +02	.26248450 +03	.19876517 +02
	.41993216 +03	.14101410 +08	.41263407 +03
1/2	.19999553 +01	.19973446 +01	.19999585 +01
	.39998011 +01	.43074675 +01	.39998156 +01
	.19998183 +02	.29953117 +02	.19997901 +02
	.41992480 +03	.47852339 +04	.42009386 +03
1/4	.19999547 +01	.19995744 +01	.19999552 +01
	.39997985 +01	.41633067 +01	.39998008 +01
	.19998160 +02	.35266650 +02	.19998181 +02
	.41992387 +03	−.22786206 +04	.41992483 +03

Diskussion: Das LAWSON-Verfahren liefert für beide Beispiele bei den betrachteten Schrittweiten keine brauchbaren Näherungswerte. Weitere Rechnungen im zweiten Beispiel, bei dem $T = f_y(x_m, u_m)$ ist, zeigen, daß LAW(4) erst bei $h = \frac{1}{8}$ für alle Komponenten die richtigen Exponenten und Vorzeichen liefert. Die mit $h = \frac{1}{32}$ ermittelte Näherungslösung besitzt allerdings in der vierten Komponente (Lösung mit stark abklingendem Anteil) für $\beta = 1$ erst eine und für $\beta = 10$ noch keine signifikante Ziffer. Die Ursache für das schlechte Verhalten des LAW(4)-Verfahrens liegt offensichtlich in dem Fehlen der S-Stabilität. Das ROS(3)-Verfahren liefert für große Schrittweiten etwas schlechtere Ergebnisse als das ARK(2)-Verfahren. Der Grund hierfür ist, wie das Experiment des ersten Beispiels (ARK(2)-Verfahren mit der Stabilitätsfunktion des ROS(3)-Verfahrens) zeigt, in dem Fehlen der L-Stabilität zu sehen.

Das LAW(4)-Verfahren benötigt in jedem Integrationsschritt vier Aufrufe von $f(x, y)$ und einen von $f_y(x, y)$. Beim ROS(3)-Verfahren sind zwei Aufrufe von $f(x, y)$, $f_x(x, y)$ und $f_y(x, y)$ pro Integrationsschritt erforderlich, während das ARK(2) zwei Aufrufe von $f(x, y)$ und einen von $f_y(x, y)$ benötigt.

Literatur

1 ALBRECHT, P., Die numerische Behandlung gewöhnlicher Differentialgleichungen, Berlin, Akademie-Verlag 1979.

2 DAHLQUIST, G., A special stability problem for linear multistep methods, BIT **3**, 27—43 (1963).

3 ENRIGHT, W. H.; HULL, T. E.; LINDBERG, B., Comparing numerical methods for stiff systems of O.D.E: s, BIT **15**, 10—48 (1975).

4 FRIEDLI, A., Verallgemeinerte RUNGE-KUTTA Verfahren zur Lösung steifer Differentialgleichungssysteme, Lecture Notes in Mathematics, Vol. 631, 35–50 (1978).
5 GRIGORIEFF, R. D., Numerik gewöhnlicher Differentialgleichungen. Bd. 1, Einschrittverfahren, Stuttgart, Teubner 1972.
6 HOUWEN, VAN DER, P. J., Construction of Integration Formulas for Initial Value Problems, Amsterdam, North-Holland Publishing Company 1976.
7 LAPIDUS, L.;SEINFELD, J. H., Numerical Solution of Ordinary Differential Equations, New York-London, Academic Press 1971.
8 LAWSON, J. D., Generalized RUNGE-KUTTA processes for stable systems with large LIPSCHITZ constants, SIAM J. Numer. Analysis 4, 372–380 (1967).
9 PROTHERO, A.; ROBINSON, A., On the stability and accuracy of one-step methods for solving stiff systems of ordinary differential equations, Math. Comput. 28, 145–162 (1974).
10 STRANG, G., Accurate partial difference methods. II: Nonlinear problems, Numer. Math. 6, 37–46 (1964).
11 VERWER, J. G., S-stability and stiff-accuracy for two classes of generalized integration method, Report NW 15/75, Mathematisch Centrum, Amsterdam 1975.
12 VERWER, J. G., Internal S-stability for generalized RUNGE-KUTTA methods, Report NW 21/75, Mathematisch Centrum Amsterdam 1975.
13 VERWER, J. G., S-stability properties for generalized RUNGE-KUTTA methods, Numer. Mathem. 27, 359–370 (1977).

Eingereicht: 4. 2. 1980, revidierte Fassung: 18. 8. 1980

Anschrift: Doz. Dr. sc. nat. KARL STREHMEL, Sektion Mathematik der Martin-Luther-Universität Halle-Wittenberg, Weinbergweg 17, DDR-4020 Halle

ZAMM 61, 261 – 264 (1981)

M. Mihelcić / K. Wingerath

$A(\alpha)$-Stable Cyclic Composite Multistep Methods of Orders 6 and 7 for Numerical Integration of Stiff Ordinary Differential Equations

Wir beschreiben zwei neue $A(\alpha)$-stabile Methoden, u. zw. eine 4-zyklische 4-Schritt und eine 5-zyklische 5-Schritt-Methode mit $P = 6$ bzw. $P = 7$. Das ist in einem gewissen Gegensatz zu den ursprünglich entwickelten Methoden [5], die einen höheren Genauigkeitsgrad ($P = 2k$), aber einen kleinen Stabilitätsbereich haben. Darüberhinaus zeigen die neuen Methoden bessere Stabilität als die Rückwärtsdifferentiationsformeln (BDF) in [6] und die zyklischen zusammengesetzten Methoden derselben Ordnung in [14].

We describe two new $A(\alpha)$-stable methods, a 4-cyclic 4-step and a 5-cyclic 5-step method with $P = 6$ and $P = 7$, respectively. This is in contrast to the originally developed methods [5] which have greater order of accuracy ($P = 2k$) but a small stability region Moreover, the new methods show better stability than the backward differentiation formulas (BDF) in [6] and th cyclic composite methods in [14] of the same order.

Описываются два нового $A(\alpha)$-устойчивого метода. Первый метод состоит из 4 цикла и 4 шага, другая из 5 циклов и 5 шагов. Это контрастирование с оригинально развивающими методами [5], которые имеют большиий порядок точности ($P = 2k$), но маленькую область устойчивости. Кроме того, новые методы показывают лучшую устойчивость чем обратными формулами дифференцирования (BDF) в [6] и циклическими составными методами самого порядка в [14].

1. Introduction

M-cyclic k-step methods (abbreviated (k, M)-methods), such as developed originally in [1], [2], [5], [13] may lead to an order of accuracy $P = 2k$ for $M = k$, without violating the condition of zero-stability. This advantage of these methods, which do not require any additional computation of the righthand side of the differential equation, is reduced, however, due to very small stability regions. The authors' experimental results have shown, e.g. that ($k = 5$, $M = 5$)-methods with $P = 10$ have stability intervals of approximately $(-2^{-15}, 0)$. In this paper we search for a ($k = 4$, $M = 4$)- and ($k = 5$, $M = 5$)-method with a large stability region.

2. Extension of stability region

The application of a (k, M)-method as a single matrix difference equation to the test equation $y' = qy$ results, with $L = h \cdot q$, in the following matrix difference equation:

$$(A_0 - LB_0)\, y_n + (A_1 - LB_1)\, y_{n+1} + \dots + (A_{j_0} - LB_{j_0})\, y_{n+j_0} = 0 \tag{1}$$

with $(M \times M)$-matrices A_i, B_i[1]) and M-vectors y_{n+i} ($i = 0, 1, \dots, j_0$) as defined in [5]. The index j_0 is defined by the condition: *if* $M \mid k$ *then* $j_0 = k/M$; *otherwise* $j_0 = [(k + M - 1)/M]$, where $[R]$ denotes the integer part of R. The *stability polynomial* of method (1) is defined by:

$$P(L, z) = \det\left(\sum_{i=0}^{j_0} (A_i - LB_i)\, z^i\right),$$

$P(L, z)$ is a polynomial of degree k in z and degree M in L. It now seems to be plausible to decrease the order $P = 2k$ of the method in order to obtain a sufficient number of free parameters to construct the polynomial $P(L, z)$ such that, for example, $A(\alpha)$-stable (k, M)-methods are possible. In [1], [10] and [12] the existence of $A(\alpha)$-stable ($k = 3$, $M = 2$)-methods with $P = 4$ and ($k = 3$, $M = 3$)-methods with $P = 5$, respectively, has been shown. Moreover, the following $A(\alpha)$-stable (k, M)-methods were found in [14]: ($k = 3$, $M = 3$) with $P = 3$, ($k = 4$, $M = 4$) with $P = 4$, ($k = 5$, $M = 4$) with $P = 5$, ($k = 6$, $M = 4$) with $P = 6$ and ($k = 7$, $M = 4$) with $P = 7$. These methods, however, do not possess the highest possible order of accuracy. In the following, we demonstrate the existence of $A(\alpha)$-stable ($k = 4$, $M = 4$)-methods with $P = 6$ and ($k = 5$, $M = 5$)-methods with $P = 7$ respectively.

3. $A(\alpha)$-stable ($k = 4$, $M = 4$)- and ($k = 5$, $M = 5$)-methods

As in [12], we demand that the stability polynomial

$$P(L, z) = \sum_{i=0}^{k} P_i(z)\, L^i \tag{2}$$

---

[1]) The matrices A_i, B_i are defined as follows:

$$A_0 = \begin{bmatrix} a_{01} & a_{11} & \dots & a_{(M-1)\,1} \\ 0 & a_{02} & \dots & a_{(M-2)\,2} \\ \vdots & & & \\ 0 & 0 & \dots & a_{0M}; \end{bmatrix}; \quad B_0 = \begin{bmatrix} b_{01} & b_{11} & \dots & b_{(M-1)\,1} \\ 0 & b_{02} & \dots & b_{(M-2)\,2} \\ \vdots & & & \\ 0 & 0 & \dots & b_{0M} \end{bmatrix} \text{etc.}$$

($P_i(z)$ are polynomials of order k) satisfies the necessary condition for $A(\alpha)$-stability, i.e. the moduli of zeros of the polynomials $P_0(z) = \lim_{L\to 0} P(L, z)$ and $P_k(z) = \lim_{L\to\infty} (P(L, z)/L^k)$ must be ≤ 1[2]). To satisfy the number of necessary conditions together with the condition for increasing the order of accuracy of the methods we need only $2k$ free parameters, so it could be possible to reach an order $P = 7$ for ($k = 4$, $M = 4$)-methods and an order $P = 9$ for ($k = 5$, $M = 5$)-methods, respectively. However this seems to be too optimistic! At any rate we were not able *to obtain these orders of accuracy.* Moreover, it seems to us that the following conjecture could be true:

"The $A(\alpha)$-stable (k, M)-methods with $k = M$ and $k > 2$ have the greatest possible order of accuracy $P = k + 2$."

However, we were unsuccessful in proving it. Following the same way as in [12] we found the new $A(\alpha)$-stable ($k = 4$, $M = 4$)-method with the accuracy of order $P = 6$ and a ($k = 5$, $M = 5$)-method with $P = 7$. The coefficients of the methods can be found in the Appendix. In addition, Table 1 displays the α- and Δ-values for the backward differentiation (BDF) methods, the cyclic composite methods [14] and the new methods. Figure 1 shows the meaning of the parameters α and Δ.

The test of the order of the new methods is shown in tables 2a and 2b by integrating the differential equation $y' = y$ for two different steplengths $h = 0{,}01$ and $h = 0{,}001$. One sees that, as the step is reduced, the absolute error decreases like h^6 and h^7, respectively.

Table 1

Order	BDF	method [14]	new method
$P = 6$			
α	17.8°	63.3°	50.5°
Δ	6.1	2.9	3.2
$P = 7$			
α	./.	33.5°	43.3°
Δ	./.	10.2	6.0

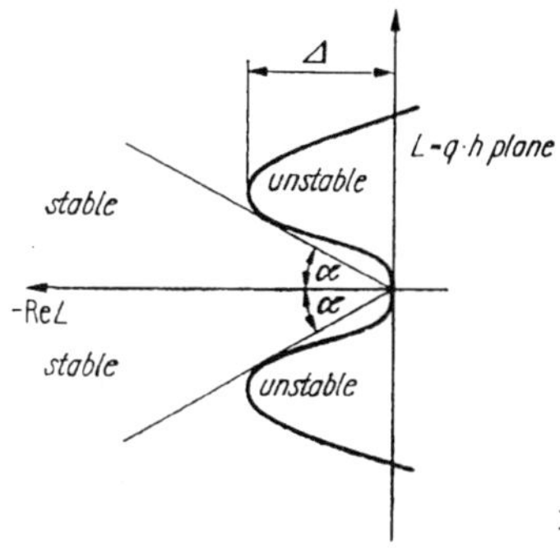

Fig. 1

Table 2a. Test of order for the method ($k = 4$, $M = 4$)

Steplength $h = 0.01$	Abs. error	Steplength $h = 0.001$	Abs. error
$x = 0.1$	$0.159 \cdot 10^{-11}$	$x = 0.1$	$0.206 \cdot 10^{-17}$
0.2	0.192 − 11	0.2	0.249 − 17
0.3	0.232 − 11	0.3	0.292 − 17
0.4	0.329 − 11	0.4	0.342 − 17
0.5	0.349 − 11	0.5	0.400 − 17

Table 2b. Test of order for the method ($k = 5$, $M = 5$)

Steplength $h = 0.01$	Abs. error	Steplength $h = 0.001$	Abs. error
$x = 0.1$	$0.157 \cdot 10^{-13}$	$x = 0.1$	$0.264 \cdot 10^{-20}$
0.2	0.272 − 13	0.2	0.317 − 20
0.3	0.330 − 13	0.3	0.379 − 20
0.4	0.393 − 13	0.4	0.450 − 20
0.5	0.467 − 13	0.5	0.532 − 20

In order to demonstrate the effectiveness of new methods the highly stiff differential equation $y' = -Q \cdot (y - x) + 1$ with $Q = 10^{12}$ was integrated with BDF, the method [14] and the new methods, respectively. The result is shown in tables 3a and 3b. The two points $x = 1{,}0$ and $x = 10{,}0$ were chosen as representative of the errors early and late in the integration. All computations were performed up to 32 decimal places. As shown in the tables, the performance of new methods were slightly poorer for $h = 0{,}01$ and $x = 1{,}0$ than the two other methods. The reason is the extremely high stiffness of the differential equation (eigenvalue $Q = 10^{12}$) and the location of the zeros of the stability polynomial $P(L, z)$ of BDF and the method [14] for $L = -\infty$ (e.g. all zeros equal zero for these methods).[3])

However, the price of this advantage is the higher stepnumber k of BDF and the method [14]. Integrating the differential equation by all of the three methods ($h = 0{,}0001$, integration interval [0, 10]), BDF and the method [14] need more computer time than the new methods (see table 4), although the differential equation is rather simple.

4. Concluding discussion

The existence of the new $A(\alpha)$-stable ($k = 4$, $M = 4$)- and ($k = 5$, $M = 5$)-methods with the accuracy of order $P = 6$ and $P = 7$, respectively, has been demonstrated. The methods show improved stability properties over the

[2]) By consistency, it is clear that one of the zeros in $P_0(z)$ must be 1.

[3]) The corresponding zeros of the new ($k = 4$, $M = 4$)- and ($k = 5$, $M = 5$)-methods are: $|z_1| = 0.271\ldots$, $|z_2| = 0.271\ldots$, $|z_3| = 0.384\ldots$, $|z_4| = 0.009$ and $|z_1| = 0.172\ldots$, $|z_2| = 0.172\ldots$, $|z_3| = 0.347\ldots$, $|z_4| = 0.347\ldots$, $|z_5| = 0.434\ldots$ (see Appendix)

Table 3a

Steplength h	x	Absolute error BDF	method [14]	new method
0.01	1.0	$0.578 \cdot 10^{-33}$	$0.173 \cdot 10^{-32}$	$0.109 \cdot 10^{-12}$
	2.0	$0.308 \cdot 10^{-32}$	$0.924 \cdot 10^{-32}$	$0.491 \cdot 10^{-27}$
	10.0	$0.308 \cdot 10^{-32}$	$0.370 \cdot 10^{-31}$	$0.154 \cdot 10^{-31}$
0.001	1.0	$0.212 \cdot 10^{-32}$	$0.289 \cdot 10^{-32}$	$0.924 \cdot 10^{-32}$
	10.0	$0.308 \cdot 10^{-32}$	$0.123 \cdot 10^{-31}$	$0.163 \cdot 10^{-30}$

Table 3b

Steplength h	x	Absolute error method [14]	new method
0.01	1.0	$0.116 \cdot 10^{-32}$	$0.539 \cdot 10^{-7}$
	2.0	$0.202 \cdot 10^{-32}$	$0.177 \cdot 10^{-14}$
	3.0	$0.308 \cdot 10^{-32}$	$0.101 \cdot 10^{-21}$
	10.0	$0.185 \cdot 10^{-31}$	$0.182 \cdot 10^{-29}$
0.001	1.0	$0.578 \cdot 10^{-33}$	$0.354 \cdot 10^{-30}$
	10.0	$0.123 \cdot 10^{-31}$	$0.317 \cdot 10^{-29}$

Table 4

Order P	BDF	Time (sec) method [14]	new method
$P = 6$	19.99	20.34	19.07
$P = 7$	./.	21.09	20.07

BDF (in case of ($k = 4$, $M = 4$)-method) and the method described in [14] (in case of ($k = 5$, $M = 5$)-method), respectively.

The existence of these methods as well as the ($k = 3$, $M = 2$)- and ($k = 3$, $M = 3$)-methods with $P = 4$ and $P = 5$, respectively, (see [10], [11] and [12]) is remarkable because, up to now, no other stiffly stable, linear methods with order greater than their stepnumber k were known.

Systematic experiments in order to obtain an $A(\alpha)$-stable ($k = 6$, $M = 6$)-method with $P = 8$ have been unsuccessful. In the author's opinion, this could be attributed to the insufficient capacity of the IBM 3033 computer because of enormous quantity of processing data.

The practical applicability of the new methods must be considered in connexion with the construction of a variable step-size, variable order integration algorithm. However, in the paper of Tendler et al. [14] a variable step-size, variable order integration algorithm is presented which is "... ideally suited for the numerical integration of stiff systems of first order ordinary differential equations $y' = f(y, t)$ wherein the Jacobian matrix $fy(y, t)$ has complex eigenvalues near the imaginary axis". This suggests that it may be possible to construct an efficient algorithm for the new methods in a similar fashion.

5. Appendix

The coefficients of the ($k = 4$, $M = 4$)-method:

A01 = −0.93903623394663787762583610050001Q − 01	B01 = −0.1853058013316520953710176130202Q − 02
A02 = 0.13137925573656420379802057858897Q + 01	B02 = −0.6428104226828121021947147465023Q + 00
A03 = 0.13943649377015345120385720850087Q + 00	B03 = −0.1234001277245579249775694759490Q + 00
A04 = 0.14441912978886368032815336562209Q + 01	B04 = −0.6918619210365557047027968203354Q + 00
A11 = 0.23077416009648853639886955190087Q + 00	B11 = 0.2451624610656680441199165441716Q + 00
A12 = −0.52185405365249169206118015478836Q + 01	B12 = 0.1217442120880404165141230674303Q + 01
A13 = −0.10116232234442301923586716889990Q + 01	B13 = 0.6120300247935214387198741562141Q + 00
A14 = −0.54921307260470900458329100837724Q + 01	B14 = 0.1168008238715629118739713853798Q + 01
A21 = 0.63627050320573670947000913374200Q + 00	B21 = −0.9774386604306546531165819680760Q + 00
A22 = 0.75294171454158511805554726478834Q + 01	B22 = 0.2924405192057778509076307429902Q + 00
A23 = 0.25338216534527796184039406821610Q + 01	B23 = −0.8155238085505402874183164469721Q + 00
A24 = 0.76604592682731494383579774102800Q + 01	B24 = 0.4386358816535462908082534556190Q + 00
A31 = −0.17731410399075614581062950756010Q + 01	B31 = 0.5535769499829904561071401752847Q + 00
A32 = −0.46246691662565762979238768858950Q + 01	B32 = −0.1439373462587944161697640697511Q + 01
A33 = −0.26616349237787028772491262016790Q + 01	B33 = −0.1145523605965158361286500199803Q − 01
A34 = −0.46125198401146961958066009827640Q + 01	B34 = −0.1467449527693897117841289393376Q + 01
A41 = 0.10000000000000000000000000000000Q + 01	B41 = 0.3644443541805902548632380173403Q + 00
A42 = 0.10000000000000000000000000000000Q + 01	B42 = 0.5385875007216307945710071168686Q + 00
A43 = 0.10000000000000000000000000000000Q + 01	B43 = 0.4094644596664487699907078389984Q + 00
A44 = 0.10000000000000000000000000000000Q + 01	B44 = 0.5438956185163976564593606928377Q + 00

Zeros of the polynomial $P(L, z)$ for $L \to -\infty$:

$$|z_1| = 0{,}271\,\ldots, \quad |z_2| = 0{,}271\,\ldots, \quad |z_3| = 0{,}384\,\ldots, \quad |z_4| = 0{,}009\,\ldots$$

The coefficients of the ($k = 5$, $M = 5$)-method:

A01 = −0.81538619050260227374341676154141Q + 00	B01 = 0.42254560713602197142718118116639Q + 00
A02 = −0.22224417606546553615938260056284Q − 01	B02 = 0.50447486432241299496034680408080Q − 01
A03 = −0.10323728835430178690862600713800Q + 00	B03 = 0.10035025395676594207675822440720Q + 00
A04 = −0.45038248555703492035752438699640Q + 00	B04 = 0.26434425320412194708011943856110Q + 00
A05 = 0.32199589112023494512610063292380Q − 02	B05 = 0.30962493792178923082813961229010Q − 01
A11 = 0.42488952540330883834818945721390Q + 01	B11 = −0.13843647398123335410907649226890Q + 01
A12 = 0.26548701214543082943049806105910Q + 00	B12 = −0.35505071357307557672195992449660Q + 00
A13 = 0.85992038163157973127806897382870Q + 00	B13 = −0.60853496713657436888081696115280Q + 00
A14 = 0.26496303820992033469567660401600Q + 01	B14 = −0.10819911309012890336494619852600Q + 01
A15 = 0.15539846605314549909060768705770Q − 01	B15 = −0.22691703684804509215104428928670Q + 00
A21 = −0.88540479513659270379005762003880Q + 01	B21 = 0.11462296725311438665691104012290Q + 01
A22 = −0.12052097458592341994150933213730Q + 01	B22 = 0.97768486590976709875548941436230Q + 00
A23 = −0.27785181722702759048028091465760Q + 01	B23 = 0.12866890843008434793797743023520Q + 01
A24 = −0.62226227266320010195981550682429Q + 01	B24 = 0.13542225211914082193786462328970Q + 01
A25 = −0.45331968167828268166798365579760Q + 00	B25 = 0.79045986222651277716152513444780Q + 00
A31 = 0.92229535519829485330980943870680Q + 01	B31 = 0.65421241909610455860990913443950Q + 00
A32 = 0.25926992769789075327023774006250Q + 01	B32 = −0.11247730127685678165079290911400Q + 01
A33 = 0.43669282009966457555417754231940Q + 01	B33 = −0.92235335158543924402497260385240Q + 00
A34 = 0.73040318879732244643062166426650Q + 01	B34 = −0.92374933116306756592146276649540Q − 01
A35 = 0.16917233449648008132611267500200Q + 01	B35 = −0.12020944686159054282112785104870Q + 01
A41 = −0.48024146641475076049359959972770Q + 01	B41 = −0.13227015739440666489380406192210Q + 01
A42 = −0.26307521256585576091018438797490Q + 01	B42 = 0.19178733809533306742638708850190Q + 00
A43 = −0.33450931220036477951084092433090Q + 01	B43 = −0.22984696536827574073333274332450Q + 00
A44 = −0.42770525181953826949239076134000Q + 01	B44 = −0.89045124549646671617297028868160Q + 00
A45 = −0.22571634688030350309534648692570Q + 01	B45 = 0.40617038177520046717524562554830Q + 00
A51 = 0.10000000000000000000000000000000Q + 01	B51 = 0.48407996565317928065364616853640Q + 00
A52 = 0.10000000000000000000000000000000Q + 01	B52 = 0.37006088463375651985204593355800Q + 00
A53 = 0.10000000000000000000000000000000Q + 01	B53 = 0.39699209789905394004672975859330Q + 00
A54 = 0.10000000000000000000000000000000Q + 01	B54 = 0.44731197571585790817249702883150Q + 00
A55 = 0.10000000000000000000000000000000Q + 01	B55 = 0.35683541060106985548535230868950Q + 00

Acknowledgement

The authors would like to thank Dr. K. Mika for helpful discussions and remarks.

References

1 Albrecht, P., Explicit, optimal stability functionals and their application to cyclic discretization methods, Computing 19, 233–249 (1978).

2 Albrecht, P., On the order of composite multistep methods for ordinary differential equations, Numer. Math. 29, 381–396 (1978).

3 Dahlquist, C., Convergence and stability in the numerical integration of ordinary differential equations. Math. Scandinav. 4, 33–53 (1956).

4 Dahlquist, C., A special stability problem for linear multistep methods. BIT 3, 27–43 (1963).

5 Donelson, J.; Hansen, E., Cyclic composite multistep predictor-corrector methods. SIAM J. Numer. Analysis 8, 137–157 (1971).

6 Gear, C. W., The automatic integration of stiff ordinary differential equations, Proc. IFIP Congress, Supplement, Booklet A, 81–85 (1968).

7 Gear, C. W., Numerical Initial Value Problems in Ordinary Differential Equations, Englewood Cliffs, N.J., Prentice-Hall 1971.

8 Henrici, P., Discrete Variable Methods in Ordinary Differential Equations, New York, J. Wiley 1962.

9 Lapidus, L.; Seinfeld, J. H., Numerical Solution of Ordinary Differential Equations, New York-London, Academic Press 1971.

10 Mihelcić, M., Fast A-stabile Donelson-Hansensche zyklische Verfahren zur numerischen Integration von stiff-Differentialgleichungssystemen, Angewandte Informatik 7, 299–305 (1977).

11 Mihelcić, M., Fast A-stabile Donelson-Hansensche zyklische Verfahren zur numerischen Integration von stiff-Differentialgleichungssystemen, Report Jül-1438, Kernforschungsanlage Jülich, Juli 1977.

12 Mihelcić, M., $A(\alpha)$-stable cyclic composite multistep methods of order 5, Computing 20, 267–272 (1978).

13 Stetter, H. J., Analysis of Discretization Methods for Ordinary Differential Equations, Berlin-Heidelberg-New York, Springer 1973.

14 Tendler, J. M.; Bickart, T. A.; Picel, Z., A stiffly stable integration process using cyclic composite methods, ACM Trans. Mathem. Software 4, 339–368 (1978).

Eingereicht: 13. 5. 1980, verbesserte Fassung 17. 10. 1980

Anschrift: Dr. M. Mihelcić, K. Wingerath, Institut für Festkörperforschung, Kernforschungsanlage Jülich, Postfach 1913, D-5170 Jülich, BRD

KLEINE MITTEILUNGEN

ZAMM 61, 265 – 267 (1981)

H. Heinrich

Eine Verallgemeinerung des arithmetisch-geometrischen Mittels

Herrn Professor Dr. F. Reutter zum siebzigsten Geburtstag

Es sei $\mathfrak{F}[I]$ die Menge aller Funktionen $f\colon I \to R$, die auf einem reellen Intervall I stetig und streng monoton sind. Mit solchen Funktionen lassen sich nach Hardy, Littlewood und Pólya [1], Kap. III, *verallgemeinerte Mittelwerte* $f^{-1}\left(\sum_{i=1}^{r} p_i f(x_i)\right)$ von r-Tupeln reeller Zahlen $x_i \in I$ bilden. Dabei sind die Zahlen p_i reelle Gewichtskoeffizienten, die der Bedingung

$$0 \leqq p_i \leqq 1\,, \qquad \sum_{i=1}^{r} p_i = 1\,,$$

genügen, und f^{-1} ist die auf dem Wertebereich $f[I]$ von f existierende und dort ebenfalls stetige und mit f gleichsinnig streng monotone Inverse von f. Dieses f-Mittel ist nach [1], Theorem 83, invariant gegenüber linearen Transformationen

$$f \to \varphi := Af + B\,, \qquad A, B \in R\,, \qquad A \neq 0;$$

das heißt, es ist

$$\varphi^{-1}\left(\sum_{i=1}^{r} p_i \varphi(x_i)\right) = f^{-1}\left(\sum_{i=1}^{r} p_i f(x_i)\right).$$

Dementsprechend erhält man mit der linearen Funktion $\varphi(x) = Ax + B$, $A \neq 0$, das mit den Gewichten p_i gewogene *arithmetische Mittel* $\sum_{i=1}^{r} p_i x_i$.

Auf dieser Grundlage wird im folgenden das Verfahren verallgemeinert, das in bekannter Weise (vgl. z. B. [2], S. 504 f.) zum Gaussschen *arithmetisch-geometrischen Mittel* eines positiven Zahlenpaares (a, b) führt. Es mögen jedoch noch einige Bemerkungen zu dem speziellen Fall $r = 2$ vorausgeschickt werden.

Es seien ein Zahlenpaar $(x_1, x_2) \in I^2$, $x_1 \neq x_2$, und ein Gewichtskoeffizient p, $0 \leqq p \leqq 1$, vorgegeben. Dann lautet das im obigen Sinn mit einer Funktion $f \in \mathfrak{F}[I]$ und den Gewichten $p_1 := 1 - p$ und $p_2 := p$ gebildete verallgemeinerte Mittel von (x_1, x_2):

$$\xi^f(x_1, x_2; p) := f^{-1}((1 - p) f(x_1) + p f(x_2))\,. \tag{1}$$

Auf Grund der einfachsten Sätze über die verallgemeinerte Mittelbildung kann ξ^f in der Form

$$\xi^f(x_1, x_2; p) = \xi(\gamma) := (1 - \gamma)\, x_1 + \gamma x_2\,, \qquad 0 \leqq \gamma \leqq 1 \tag{2}$$

angesetzt werden, und es gelten die Äquivalenzen

$$p = 0 \quad \Leftrightarrow \gamma = 0\,, \tag{2.1}$$

$$p = 1 \quad \Leftrightarrow \gamma = 1\,, \tag{2.2}$$

$$0 < p < 1 \Leftrightarrow 0 < \gamma < 1\,. \tag{2.3}$$

Ist φ eine lineare Funktion: $\varphi(x) = Ax + B$, $A \neq 0$, so ist $\gamma = p$. Es ist dann

$$\xi^\varphi(x_1, x_2; p) = \xi(p) := (1 - p)\, x_1 + p x_2 \tag{3.1}$$

das *zum Gewichtskoeffizienten p gehörige arithmetische Mittel* von (x_1, x_2).

Im Fall (2.3) heißt die Mittelbildung „echt.“ ξ^f liegt dann zwischen x_1 und x_2. Bild 1 veranschaulicht diesen Fall. In einem x,y-Koordinatensystem liege die Funktionskurve $y = f(x)$ vor. $\Pi(p)$ sei derjenige Punkt der Kurvensehne P_1P_2 zwischen x_1 und x_2, der das arithmetische Mittel $\xi(p)$ als Abszisse hat. Dann ist seine Ordinate

$$\eta(p) = (1 - p)\, f(x_1) + p f(x_2)\,. \tag{3.2}$$

Das ist aber nach (1) gleich $f(\xi^f)$. Daher ist das verallgemeinerte Mittel (1) die Abszisse desjenigen eindeutig bestimmten Punktes Π^f der Funktionskurve, der dieselbe Ordinate wie der Punkt $\Pi(p)$ hat. Mit dem Streckenzug, der von der Stelle $\xi(p)$ der x-Achse über die Punkte $\Pi(p)$ und Π^f zur Stelle ξ^f der x-Achse führt, ist somit eine einfache Konstruktion des verallgemeinerten Mittels ξ^f gegeben.

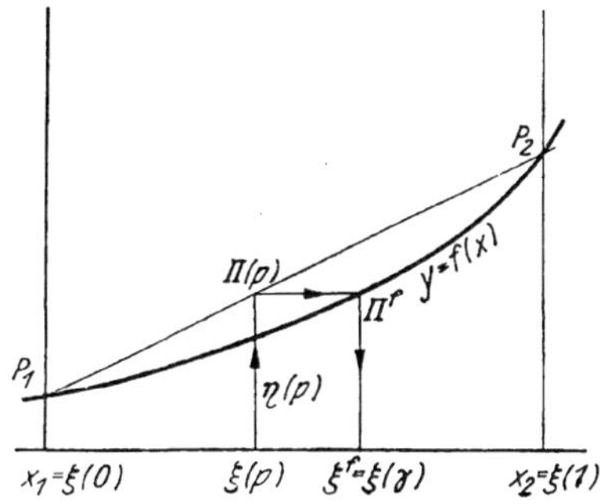

Bild 1

Weiter unten werden die Steigungen (dividierten Differenzen) 1. und 2. Ordnung der Funktion f zwischen den Punkten x_i und ξ^f bzw. x_1, x_2 und ξ^f benötigt. In der in [3], S. 134, verwendeten Schreibweise sind sie durch

$$[f; x_i, \xi^f] := \frac{f(x_i) - f(\xi^f)}{x_i - \xi^f}\,, \qquad i = 1, 2\,, \tag{4.1}$$

$$[f; x_1, x_2, \xi^f] := \frac{[f; x_1, \xi^f] - [f; x_2, \xi^f]}{x_1 - x_2} \tag{4.2}$$

definiert. Mit (1) und (2) ergeben sich daraus die Werte

$$[f; x_1, \xi^f] = \frac{p}{\gamma}\,[f; x_1, x_2]\,,$$

$$[f; x_2, \xi^f] = \frac{1 - p}{1 - \gamma}\,[f; x_1, x_2]\,, \tag{5.1}$$

$$[f; x_1, x_2, \xi^f] = \frac{p - \gamma}{\gamma(1 - \gamma)}\,\frac{[f; x_1, x_2]}{x_1 - x_2}\,. \tag{5.2}$$

Wegen der strengen Monotonie von f ist dabei $[f; x_1, x_2] := (f(x_1) - f(x_2))/(x_1 - x_2) \neq 0$. Damit lassen sich die Steigungsverhältnisse

$$\sigma_i^f(x_1, x_2; p) := \frac{[f; x_1, x_2, \xi^f]}{[f; x_i, \xi^f]} = \frac{p - \gamma}{p(1 - \gamma)} \cdot \frac{1}{x_1 - x_2} \qquad \text{für } i = 1$$

$$\text{bzw.} \quad = \frac{p - \gamma}{\gamma(1 - p)} \cdot \frac{1}{x_1 - x_2} \qquad \text{für } i = 2 \tag{6.1}$$

bilden. (Daß in diesen Relationen für die Zahl γ, die die Lage von ξ^f zwischen x_1 und x_2 bestimmt, die Steigungsverhältnisse σ_i^f vorkommen, steht in engem Zusammenhang mit der eingangs erwähnten Invarianz des verallgemeinerten Mittels gegenüber linearen Transformationen der verwendeten Funktion. Denn auch σ_i^f weist wegen der Linearität der Steigungsoperation (vgl. [3], S. 140f.) diese Invarianz auf.)

Aus (6.1) läßt sich u. a. auch das Vorzeichen von $\gamma - p$ ablesen, das die Richtung der Verschiebung vom arithmetischen Mittel $\xi(p)$ zum f-Mittel $\xi^f = \xi(\gamma)$ charakterisiert. Wegen

$$p(1 - \gamma) > 0 \text{ bzw. } \gamma(1 - p) > 0$$

ist

$$\operatorname{sgn}(p - \gamma) = \operatorname{sgn}[(x_1 - x_2)\, \sigma_i^f(x_1, x_2; p)]\,. \tag{6.2}$$

Ist beispielsweise die Funktion f zwischen x_1 und $x_2 > x_1$ monoton wachsend und konvex (vgl. Bild) oder monoton fallend und konkav, so haben in (6.1) die erste und die zweite Steigung dasselbe Vorzeichen, und es ist daher $\sigma_i^f > 0$, $p - \gamma < 0$.

Für das eingangs angekündigte Verfahren werden nunmehr folgende Vorgaben benötigt:

V_1: ein reelles Intervall I;

V_2: zwei Funktionen $f, g \in \mathfrak{F}[I]$ mit den Inversen f^{-1} und g^{-1};

V_3: ein Gewichtskoeffizient p mit $0 < p < 1$;

V_4: ein reelles Zahlenpaar (a, b) mit $a \neq b$.

Das Verfahren beginnt dann mit

$$u_0 := a\,, \qquad v_0 := b \tag{7.1}$$

und wird für $k = 0, 1, 2, \ldots$ fortgesetzt mit

$$u_{k+1} := \xi^f(u_k, v_k; p) = f^{-1}((1-p)\,f(u_k) + pf(v_k))\,, \tag{7.2.1}$$

$$v_{k+1} := \xi^g(u_k, v_k; p) = g^{-1}((1-p)\,g(u_k) + pg(v_k))\,. \tag{7.2.2}$$

Die in V_3 vorgenommene Beschränkung auf echte Mittelbildungen ist dadurch gerechtfertigt, daß das Verfahren für $p = 0$ und für $p = 1$ beim Ausgangswert a bzw. b stehenbleibt und daher uninteressant ist. Denn es gilt für $p = 0$:

$$u_1 = f^{-1}(f(u_0)) = u_0 = a\,, \qquad v_1 = g^{-1}(g(u_0)) = u_0 = a\,,$$

d. h. $u_1 = v_1 = a$ und damit auch weiterhin $u_k = v_k = \ldots = u_0 = a$. Ähnlich schließt man für $p = 1$.

Für $0 < p < 1$ gilt nun der

Satz 1: *Das durch (7.1) und (7.2.1/2) festgelegte Verfahren konvergiert stets — unter Umständen bereits nach endlich vielen Schritten — als ein Einschließungsverfahren gegen einen gemeinsamen Grenzwert*

$$a^* := M^{f,g}(a, b; p) \tag{8}$$

der beiden Zahlenfolgen $\{u_k\}$ und $\{v_k\}$, das zum Gewichtskoeffizienten p gehörige (f, g)-Grenzmittel des Zahlenpaares (a, b).

Beweis: Zum Beweis werden die Werte

$$a_k := \min\{u_k, v_k\}\,, \qquad b_k := \max\{u_k, v_k\} \tag{9.1}$$

eingeführt. Dann gilt bei jedem Schritt wegen V_3:

$$a_k < a_{k+1} < b_k \quad \text{und} \quad a_k < b_{k+1} < b_k\,,$$

insgesamt also

$$a_k < a_{k+1} \leqq b_{k+1} < b_k\,. \tag{9.2}$$

Folgende beiden Fälle sind zu unterscheiden:

1) Es gibt einen Index $n > 0$, für den erstmals $u_n = v_n (= a_n = b_n)$ ist. Dann liefert das Verfahren für $k = n$ und anschließend für jedes weitere $k > n$

$$u_{k+1} = v_{k+1} = a_n; \tag{10.1}$$

das heißt, es bleibt bei dem Wert

$$a^* := a_n \tag{10.2}$$

stehen.

2) Es gibt keinen Index n, für den $u_n = v_n$ ist. Dann gilt nach (9.2) für alle $k = 0, 1, 2, \ldots$

$$a_0 < a_1 < \ldots < a_k < b_k < \ldots < b_1 < b_0\,. \tag{11.1}$$

Es konvergieren aber die monoton wachsende Folge $\{a_k\}$ und die monoton fallende Folge $\{b_k\}$, da sie beide in Monotonierichtung beschränkt sind, gegen Grenzwerte a^* und b^*, für die

$$a^* \leqq b^* \tag{11.2}$$

gilt. Es ist zu zeigen, daß die Annahme $a^* < b^*$ auf einen Widerspruch führt und daher stets $a^* = b^*$ gilt.

Für den Nachweis genügt es, die Eigenschaften von f (oder g) allein heranzuziehen. Aus $a^* < b^*$ folgt wegen der strengen Monotonie von f: $f(a^*) \neq f(b^*)$. Daher läßt sich zu jedem festen $p \in (0, 1)$ ein $\varepsilon > 0$ wählen, so daß

$$0 < \varepsilon \leqq \min\left\{\frac{p}{1-p}\,|f(a^*) - f(b^*)|\,, \quad \frac{1-p}{p}\,|f(a^*) - f(b^*)|\right\} \tag{11.3}$$

ist. Zu jedem solchen ε gibt es wegen der Stetigkeit und Monotonie von f ein $\alpha(\varepsilon)$ und ein $\beta(\varepsilon)$, so daß

$$|f(a^*) - f(a^* - \alpha)| < \varepsilon \quad \text{und} \quad |f(b^* + \beta) - f(b^*)| < \varepsilon \tag{11.4}$$

ist. Des weiteren gibt es wegen der Konvergenz $a_k \nearrow a^*$ und $b_k \searrow b^*$ zu jedem $\alpha > 0$ bzw. $\beta > 0$ einen Index $n(\alpha)$ bzw. $n(\beta)$, so daß für alle $k > n(\alpha)$ bzw. $k > n(\beta)$

$$a^* - \alpha < a_k < a^* \quad \text{bzw.} \quad b^* < b_k < b^* + \beta \tag{11.5}$$

gilt. Es sei $n_0 := \max\{n(\alpha), n(\beta)\}$. Die Indexmenge

$$N_0 := \{k \in N;\ k > n_0\} \tag{11.6}$$

werde in die Teilmengen

$$N_1 := \{k \in N_0;\ u_k < v_k \Rightarrow a_k = u_k,\ b_k = v_k\}\,, \tag{11.6.1}$$

$$N_2 := \{k \in N_0;\ u_k > v_k \Rightarrow a_k = v_k,\ b_k = u_k\} \tag{11.6.2}$$

zerlegt, von denen eine (aber nur eine) auch leer sein kann. Dann sind vier Fälle möglich, je nachdem ob f monoton steigt ($\nearrow$) oder fällt ($\searrow$) und ob der ausgewählte Index k zu N_1 oder zu N_2 gehört. Es genügt, einen dieser Fälle zu behandeln, beispielsweise $f\nearrow$, $k \in N_1$. Nach (11.5) und (11.6.1) ist dann für alle $k > n_0$

$$f(a^* - \alpha) < f(u_k) < f(a^*)\,, \tag{11.7.1}$$

$$f(b^*) < f(v_k) < f(b^* + \beta)\,. \tag{11.7.2}$$

Durch Linearkombination mit den Koeffizienten $1 - p > 0$ und $p > 0$ folgt daraus mit (7.2.1)

$$(1-p)\,f(a^* - \alpha) + pf(b^*) < f(u_{k+1}) < (1-p)\,f(a^*) + pf(b^* + \beta)\,. \tag{11.8}$$

Nun gilt nach (11.3) und (11.4)

$$f(a^*) - f(a^* - \alpha) < \varepsilon \leqq \frac{p}{1-p}\,(f(b^*) - f(a^*))\,, \tag{11.8.1}$$

$$f(b^* + \beta) - f(b^*) < \varepsilon \leqq \frac{1-p}{p}\,(f(b^*) - f(a^*))\,; \tag{11.8.2}$$

das ist aber gleichbedeutend mit

$$(1-p)\,f(a^* - \alpha) + pf(b^*) < f(a^*)\,, \tag{11.9.1}$$

$$(1-p)\,f(a^*) + pf(b^* + \beta) < f(b^*)\,.$$

Setzt man dies in (11.8) ein und wendet $f^{-1}\nearrow$ an, so ergibt sich

$$f(a^*) < f(u_{k+1}) < f(b^*) \Rightarrow a^* < u_{k+1} < b^*\,. \tag{11.10}$$

Für $k + 1 \in N_1$ steht dies im Widerspruch zu $u_{k+1} = a_{k+1} < a^*$, für $k + 1 \in N_2$ im Widerspruch zu $u_{k+1} = b_{k+1} > b^*$. Damit ist nachgewiesen, daß $a^* = b^*$ ist, d. h.

$$\lim_{k\to\infty} |u_k - v_k| = \lim_{k\to\infty} (b_k - a_k) = 0\,.$$

Das Verfahren liefert danach eine Intervallschachtelung für a^* als den gemeinsamen Grenzwert der Folgen $\{u_k\}$ und $\{v_k\}$.

Beispiele:

1. Mit $f(x) = \ln x$, $g(x) = x$, $x > 0$ und $p = \frac{1}{2}$ erhält man für das Zahlenpaar (a, b), $0 < a < b$, das klassische Verfahren

$$u_0 := a\,, \qquad v_0 := b$$

$$u_{k+1} := \sqrt{u_k v_k}\,, \qquad v_{k+1} := \tfrac{1}{2}\,(u_k + v_k)\,, \qquad k = 0, 1, 2, \ldots$$

das als Grenzwert a^* das Gausssche *arithmetisch-geometrische Mittel* agM(a, b) liefert.

2. Die zu entgegengesetzten Exponenten $s \neq 0$ und $-s$ gehörigen Potenzen $f(x) = x^s$, $g(x) = x^{-s}$, $x < 0$, liefern mit dem Gewichtskoeffizienten $p = 1/2$ vom positiven Zahlenpaar $(u_0, v_0) := (a, b)$ ausgehend das Verfahren

$$u_{k+1} := \left(\tfrac{1}{2}(u_k^s + v_k^s)\right)^{1/s}\,, \qquad v_{k+1} := \left(\tfrac{1}{2}\,(u_k^{-s} + v_k^{-s})\right)^{-1/s}.$$

Wie sich leicht bestätigen läßt, gilt

$$u_{k+1}v_{k+1} = u_k v_k = \ldots = u_0 v_0 = ab\,,$$

so daß sich als Grenzwert a^* das geometrische Mittel $g(a, b) = \sqrt{ab}$ ergibt. Für $s = 1$ wäre dieser Grenzwert in Analogie zum Beispiel 1 als das *harmonisch-arithmetische Mittel* von (a, b) zu bezeichnen.

3. Es sei $f(x) = x^2$, $g(x) = e^{-x}$, $x > 0$; mit $p = 0.3$ lautet das Verfahren:

$$u_{k+1} := \sqrt{0.7u_k^2 + 0.3v_k^2}$$

$$v_{k+1} := -\ln\,(0.7\,e^{-u_k} + 0.3\,e^{-v_k})\,.$$

Von $(u_0, v_0) := (a, b) = (0.1, 2)$ ausgehend erhält man damit die ne nstehenden Zahlenwerte.

k	u_k	v_k
0	0.1	2
1	1.0986355	0.39454479
2	0.94424555	0.83120549
3	0.91180619	0.90897201
4	0.91095686	0.91095509
5	0.91095633	0.91095633

Von dem klassischen Beispiel 1 ist bekannt, daß das Verfahren quadratisch konvergiert (s. z. B. [4], S. 418f.). Dies trifft nun auch für das verallgemeinerte Verfahren zu, sofern es nicht abbricht. Es gilt der

Satz 2: *Sind die zu den Mittelwerten $\xi_k^f = u_{k+1}$ und $\xi_k^g = v_{k+1}$ gehörigen Steigungsverhältnisse*

$$\sigma_k^f := \frac{[f; u_k, v_k, u_{k+1}]}{[f; u_k, u_{k+1}]}, \qquad \sigma_k^g := \frac{[g; u_k, v_k, v_{k+1}]}{[g; v_k, v_{k+1}]}$$

dem Betrage nach gleichmäßig beschränkt und $u_k \neq v_k$ für alle k, so gibt es eine Konstante $C \geqq 0$, so daß für $k > n_0$

$$|u_{k+1} - v_{k+1}| \leqq C(u_k - v_k)^2$$

ist. In diesem Sinne konvergiert das in Rede stehende Einschließungsverfahren mindestens quadratisch gegen das (f, g)-Grenzmittel

$$a^* = M^{f,g}(a, b; p).$$

Beweis: Mit den Ansätzen — vgl. (2) —

$$u_{k+1} = (1 - \alpha_k)\, u_k + \alpha_k v_k\,, \qquad 0 < \alpha_k < 1 \qquad (12.1.1)$$

$$v_{k+1} = (1 - \beta_k)\, u_k + \beta_k v_k\,, \qquad 0 < \beta_k < 1 \qquad (12.1.2)$$

ergeben sich im Anschluß an (6.1) für σ_k^f und σ_k^g die Ausdrücke

$$\sigma_k^f = \frac{p - \alpha_k}{(1 - \alpha_k)\, p} \cdot \frac{1}{u_k - v_k}\,, \qquad \sigma_k^g = \frac{p - \beta_k}{\beta_k(1 - p)} \cdot \frac{1}{u_k - v_k}\,. \qquad (12.2)$$

Damit wird

$$\frac{u_{k+1} - v_{k+1}}{(u_k - v_k)^2} = \frac{\beta_k - \alpha_k}{u_k - v_k} = \frac{p - \alpha_k}{u_k - v_k} - \frac{p - \beta_k}{u_k - v_k} =$$

$$= (1 - \alpha_k)\, p\sigma_k^f - \beta_k(1 - p)\, \sigma_k^g\,. \qquad (13)$$

Dieser Ausdruck ist aber wegen $0 < (1 - \alpha_k)\, p < 1$ und $0 < \beta_k(1 - p) < 1$ zugleich mit σ_k^f und σ_k^g dem absoluten Betrage nach gleichmäßig beschränkt. Damit ist die Behauptung bewiesen.

Sind die Funktionen f und $g \in \mathfrak{F}[I]$ — wenigstens in einer hinreichend großen Umgebung von a^* zweimal stetig differenzierbar und $f'(x) \neq 0$ und $g'(x) \neq 0$ für alle $x \in I$, so ist die gleichmäßige Beschränktheit von $|\sigma_k^f|$ und $|\sigma_k^g|$ wegen des engen Zusammenhanges zwischen den Steigungen und den Ableitungen gleicher Ordnung (vgl. [3], S. 139). Führt man die Funktionen

$$\sigma^f(x) := \frac{f''(x)}{f'(x)}\,, \qquad \sigma^g(x) := \frac{g''(x)}{g'(x)} \qquad (14.1)$$

ein und berücksichtigt man, daß, wie sich aus (6.1) mit $\gamma = \alpha_k$ bzw. $= \beta_k$, $x_1 = u_k$, $x_2 = v_k$ für $u_k - v_k \to 0$ leicht herleiten läßt,

$$\lim_{k\to\infty} \alpha_k = \lim_{k\to\infty} \beta_k = p \qquad (14.2)$$

ist, so ergibt sich für die rechte Seite von (13) der Grenzwert

$$\lim_{k\to\infty} \frac{u_{k+1} - v_{k+1}}{(u_k - v_k)^2} = p(1 - p)\, (\sigma^f(a^*) - \sigma^g(a^*))\,. \qquad (14.3)$$

Außer dem Grenzwert a^* ändert sich an den Konvergenzeigenschaften des Verfahrens nichts, wenn man es statt mit einem festen Gewichtskoeffizienten p mit einer Folge von Gewichtskoeffizienten p_k durchführt, sofern diese der Bedingung $0 < p' \leqq p_k \leqq p'' < 1$ genügen.

Der vom jungen GAUSS entdeckte Zusammenhang zwischen dem elliptischen Integral $\int\limits_0^1 \frac{1}{\sqrt{1 - x^4}}\, dx$ und dem agM $(1, \sqrt{2})$ — s. seine berühmte Tagebuchnotiz Nr. 98 vom 30. Mai 1799 — läßt es denkbar erscheinen, daß es Probleme gibt, bei deren numerischer Lösung sich auch das verallgemeinerte Verfahren dank seiner Eigenschaften als quadratisch konvergierendes Einschließungsverfahren mit Erfolg einsetzen ließe.

Literatur

1 HARDY, G. H.; LITTLEWOOD, J. E.; PÓLYA, G., Inequalities (2. Aufl.), Cambridge 1962.
2 STRUBECKER, K., Einführung in die Höhere Mathematik. Bd. I, München-Wien 1966.
3 HEINRICH, H., Einführung in die Praktische Analysis. Bd. I, Leipzig-Aachen 1963.
4 BULIRSCH, R.; STOER, J., Darstellung von Funktionen in Rechenautomaten, in: SAUER, R.; SZABÓ, I., Mathematische Hilfsmittel des Ingenieurs. Bd. III, Berlin-Heidelberg-New York 1968.
5 ZURMÜHL, R., Praktische Mathematik für Ingenieure und Physiker (5. Aufl.), Berlin-Heidelberg-New York 1965.

Eingereicht: 14. 10. 1980

Anschrift: Prof. (em.) Dr. Dr. h. c. HELMUT HEINRICH, Friedrich-Hegel-Straße 31, DDR-8027 Dresden

ZAMM 61, 267–269 (1981)

C. Y. WANG

Equilibrium of a Heavy, Naturally Curved Sheet on an Inclined Plane

1. Introduction and formulation

Thin sheets of elastic material (metal, plastic, cloth etc.) sometimes have a natural curvature due to processing or due to previous storage on circular drums. When the material is unrolled from the drum and laid on a plane, an equilibrium shape may exist due to the balance of gravity and the sheet's tendency to roll up again. The present note studies the properties of these equilibrium shapes.

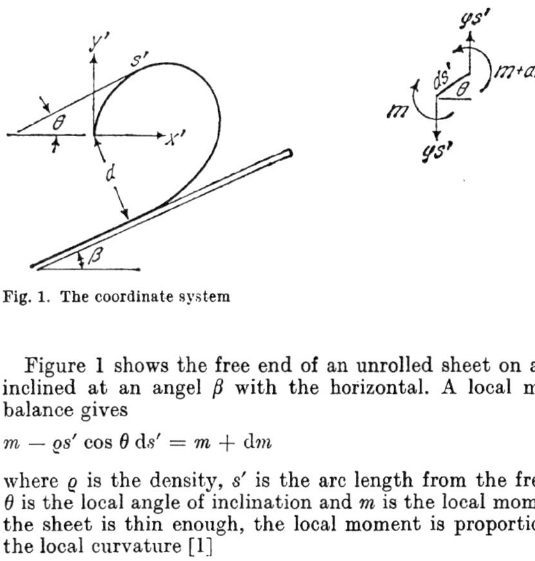

Fig. 1. The coordinate system

Figure 1 shows the free end of an unrolled sheet on a plane inclined at an angel β with the horizontal. A local moment balance gives

$$m - \varrho s' \cos\theta\, \mathrm{d}s' = m + \mathrm{d}m \qquad (1)$$

where ϱ is the density, s' is the arc length from the free end, θ is the local angle of inclination and m is the local moment. If the sheet is thin enough, the local moment is proportional to the local curvature [1]

$$m = EI\, \frac{\mathrm{d}\theta}{\mathrm{d}s'}\,. \qquad (2)$$

Here EI is the flexural rigidity. Equations (1, 2) give

$$EI\, \frac{\mathrm{d}^2\theta}{\mathrm{d}s'^2} = -\varrho s' \cos\theta\,. \qquad (3)$$

We normalize the length as follows

$$s = \frac{s'}{(EI/\varrho)^{1/3}}\,. \qquad (4)$$

Equation 3 becomes

$$\frac{\mathrm{d}^2\theta}{\mathrm{d}s^2} = -s\cos\theta\,. \qquad (5)$$

One boundary condition is

$$\frac{d\theta}{ds}(0) = -A\,, \tag{6}$$

where

$$A \equiv \frac{(EI/\varrho)^{1/3}}{R} \tag{7}$$

denotes the relative importance of EI, ϱ, and the natural radius of curvature R. The other boundary condition is at the place where the sheet separates from the incline, or when

$$\theta = \beta - \pi\,, \qquad \frac{d\theta}{ds} = 0\,. \tag{8}$$

The shape of the sheet can be obtained by integrating for the normalized cartesian coordinates x, y:

$$\frac{dx}{ds} = \cos\theta\,, \qquad \frac{dy}{ds} = \sin\theta\,, \tag{9}$$

$$x(0) = y(0) = 0\,. \tag{10}$$

Equations (5, 6, 8—10) do not have closed form solutions.

2. Approximate solution for small $A, \beta \neq \pm 90°$

Small A means the sheet is relatively compliant or the density is large or natural curvature is small. We expect the sheet lie very close to the inclined plane. Set

$$A \equiv \varepsilon^2 \ll 1\,, \tag{11}$$

$$s = \varepsilon^\delta t\,, \qquad \theta = \beta - \pi + \varepsilon^{\gamma_1}\theta_1(t) + \varepsilon^{\gamma_2}\theta_2(t) + \varepsilon^{\gamma_3}\theta_3(t) + \dots \tag{12}$$

Substitution of equations (11, 12) into equations (5, 6) yields the appropriate stretching powers $\delta = 1$, $\gamma_n = 3n$, provided $\beta \neq \pm 90°$. The successive equations to be solved are

$$\frac{d^2\theta_1}{dt^2} = t\cos\beta\,, \qquad \frac{d^2\theta_2}{dt^2} = -t\theta_1 \sin\beta\,, \tag{13), (14}$$

$$\frac{d\theta_1}{dt}(0) = 1\,, \qquad \frac{d\theta_2}{dt}(0) = 0\,. \tag{15}$$

The boundary condition equation (8) is more difficult since we do not know where it occurs. Let this location be at $s = s^*$,

$$s^* = \varepsilon(t_0 + \varepsilon^3 t_1 + \dots)\,. \tag{16}$$

Using equation (16), we expand equation (8) in a TAYLOR series about $t = t_0$. The result is

$$\theta_1(t_0) = 0\,, \qquad \frac{d\theta_1}{dt}(t_0) = 0\,, \tag{17}$$

$$\theta_2(t_0) = 0\,, \qquad \frac{d\theta_2}{dt}(t_0) + t_1\frac{d^2\theta_1}{dt^2}(t_0) = 0\,. \tag{18}$$

Thus the first order solution is found to be

$$\theta_1 = \frac{\cos\beta}{6}t^3 - t + \frac{2}{3}\sqrt{\frac{2}{\cos\beta}}\,, \tag{19}$$

$$t_0 = \sqrt{\frac{2}{\cos\beta}}\,. \tag{20}$$

The second order solution is

$$\theta_2 = -\sin\beta\left(\frac{\cos\beta}{180}t^6 - \frac{1}{12}t^4 + \frac{1}{9}\sqrt{\frac{2}{\cos\beta}}\,t^3 - \frac{7}{45\cos^2\beta}\right), \tag{21}$$

$$t_1 = \frac{2\sin\beta}{15\cos^2\beta}\,. \tag{22}$$

Using equations (9, 10), the shape is

$$x = -\varepsilon\cos\beta t + \varepsilon^4 \sin\beta\left(\frac{\cos\beta}{24}t^4 - \frac{1}{2}t^2 + \frac{2}{3}\sqrt{\frac{2}{\cos\beta}}\,t\right) + O(\varepsilon^7)\,, \tag{23}$$

$$y = -\varepsilon\sin\beta t - \varepsilon^4\cos\beta\left(\frac{\cos\beta}{24}t^4 - \frac{1}{2}t^2 + \frac{2}{3}\sqrt{\frac{2}{\cos\beta}}\,t\right) + O(\varepsilon^7)\,. \tag{24}$$

Of interest is the distance from the edge of the sheet to the incline

$$d = (-y\cos\beta + x\sin\beta)|_{s=s^*}\,. \tag{25}$$

Using equations (16. 20, 22—24) we find

$$d = \varepsilon^4\frac{1}{2\cos\beta} + \sigma(\varepsilon^7)\,. \tag{26}$$

It is easy to show that θ_3 is even in β and the $O(\varepsilon^7)$ term of equation (26) is odd in β. Thus for $\beta = 0$

$$d = \frac{1}{2}\varepsilon^4 + O(\varepsilon^{11})\,. \tag{27}$$

3. Approximate solution for small $A, \beta = -90°$

When $\beta \geq 90°$ the sheet of course will not be in equilibrium for any A. When $\beta = -\pi/2$ the sheet hangs vertically down with $s^* = \infty$. Our solution in the previous section is not valid for β close to $-90°$. Using expansions similar to equations (11, 12), we find, for $\beta = -\pi/2$,

$$\theta = -3\pi/2 + \varepsilon^2\varphi_1(s) + \varepsilon^6\varphi_2(s) + \dots \tag{28}$$

The leading equation is

$$\frac{d^2\varphi_1}{ds^2} = s\varphi_1\,, \tag{29}$$

$$\frac{d\varphi_1}{ds}(0) = -1\,, \qquad \varphi_1(\infty) = 0\,. \tag{30}$$

The solution is the AIRY function [2]

$$\varphi_1 = 3^{1/3}\Gamma\left(\frac{1}{3}\right)Ai(s)\,. \tag{31}$$

Using equations (9, 10, 27, 31) the shape is found to be

$$x = -\varepsilon^2\int_0^s \varphi_1(s)\,ds + O(\varepsilon^6)\,, \tag{32}$$

$$y = s - \frac{\varepsilon^4}{2}\int_0^s \varphi_1^2(s)\,ds + O(\varepsilon^8)\,. \tag{33}$$

The distance from the edge of the sheet to the incline is

$$d = \varepsilon^2\int_0^\infty \varphi_1(s)\,ds + O(\varepsilon^6)$$

$$= \varepsilon^2 3^{-2/3}\Gamma\left(\frac{1}{3}\right) + O(\varepsilon^6) = 1.2879\varepsilon^2 + O(\varepsilon^6)\,. \tag{34}$$

4. Numerical solution

For large A, numerical solution of equations (5, 6, 8—10) is necessary. The method is to take any $\theta(0)$ and integrate equations (5, 6, 9, 10) as an initial value problem by the fourth order RUNGE-KUTTA algorithm until $d\theta/ds = 0$. We thus obtain s^* and $\beta = \theta(s^*) + \pi$. The distance d is then found from equation (25).

Figure 2 shows the distance d as a function of β for each constant A. Let us concentrate the curved sheet with $A = 1$.

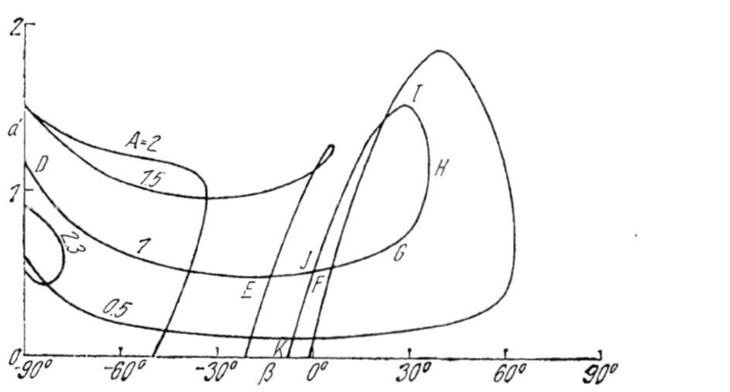

Fig. 2. Normalized distance d as a function of β for various A values

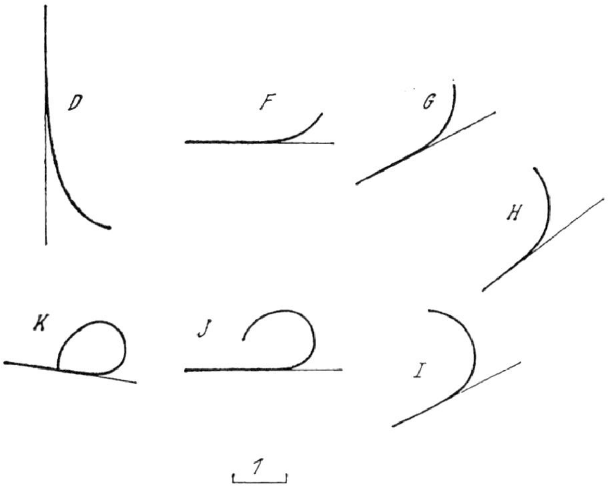

Fig. 3. Predicted equilibrium shapes for $A = 1$. Letters correspond to states shown in Figure 2

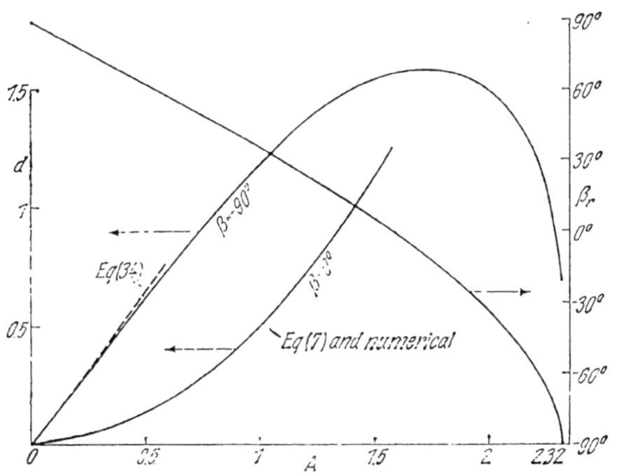

Fig. 4. Roll-up angle β_r and distance d as a function of A

As β is increased from $-90°$ (state D, the vertical hanging position) d decreases to a minimum at state E and then increases until state H is reached at $\beta = 36.5°$. There exists no equilibrium shape for $\beta > 36.5°$, indicating dynamic roll up of the sheet occurs. If β is decreased from state H, one of two possibilities may happen. The equilibrium states would either reverse itself by following HGFE or follow the path HIJK. It is seen that nonuniqueness occurs when $-8° \leqq \beta \leqq 36.5°$. Figure 3 depicts some of the shapes. Notice states G and I are different solutions for the same $\beta = 27°$. Notice also that although states F and J are different solutions for $\beta = 0$, the distance d is the same.

Figure 4 shows the roll-up angle as a function of A. If $A \geqq 2.32$, roll-up always occurs, i.e. the sheet cannot be unrolled without outside restraining forces. Also shown in Figure 4 is the edge distance d for $\beta = -90°$ and $\beta = 0°$. For both inclination angles d is unique except perhaps for values of A close to 2.32. The curve for $\beta = 0°$ terminates at $A = 1.58$ due to roll-up. Our approximate solution, equation (34), agrees well with an exact numerical solution for $A < 0.5$. Our approximation solution equation (7), surprisingly, is almost identical with our numerical results.

Figure 4 is also useful in the industrial testing of the elastic properties of curved sheets. (Reference [3] gives testing methods for straight sheets). One can easily measure the density ϱ, the roll-up angle β_r and the dimensional distance d' (if laid horizontally, $\beta = 0$, or if hanging vertically, $\beta = -90°$). Since

$$d = d'(EI/\varrho)^{-1/3}, \qquad A = (IE/\varrho)^{1/3}/R\,. \tag{35}$$

One can use Figure 4 to obtain inversely the flexural rigidity EI and natural curvature R.

References

1 TIMOSHENKO, S. P., and GERE, J. M., Theory of Elastic Stability, McGraw-Hill, New York 1961, Chap. 2.
2 ABRAMOWITZ, M., and STEGUN, I. A., Eds., Handbook of Mathematical Functions, Dover, New York 1965, Chap. 10.
3 BICKLEY, W. G., The heavy elastica, Phil. Mag. Ser. 7, Vol. **17**, 1934, pp. 603 to 622.

Eingereicht am 15. 9. 1980

Anschrift: Prof. C. Y. WANG, Michigan State University, Department of Mathematics, Wells Hall, East Lansing, Michigan, 48824, USA

ZAMM 61, 269 – 272 (1981)

O. BECKER

Zur Berechnung von Integralen über die Balkenfunktionen

1. Einleitung

Probleme der Elastomechanik lassen sich mit vertretbarem Aufwand häufig nur näherungsweise lösen, indem z. B. mit Hilfe des Verfahrens von RITZ oder GALERKIN das elastische Kontinuum in ein Ersatzsystem mit n Freiheitsgraden überführt wird. Jeder der n Summanden eines entsprechenden Näherungsansatzes wird dabei als Produkt einer bekannten Ortsfunktion und einer unbekannten Zeitfunktion bzw. Konstanten dargestellt. Durch Integration über den Definitionsbereich der Ortsfunktionen erhält man n gewöhnliche Differentialgleichungen bzw. Gleichungen, die das Verhalten des Ersatzsystems beschreiben.

Für Stabsysteme und bestimmte Klassen von Flächentragwerken werden als Ortsfunktionen vorteilhaft die sog. Balkenfunktionen verwendet, die als Eigenfunktionen des schwingenden Biegebalkens ein vollständiges, orthogonales Funktionensystem darstellen. Die hierbei auftretenden Integrale über je zwei Balkenfunktionen bzw. deren Ableitungen werden in der Literatur jeweils nur für ganz bestimmte Lagerungen angegeben und sie beschränken sich i. a. nur auf den Fall, daß die Summe der Ableitungen null, zwei oder vier ist [1], [2], [3], [4]. Bei gekrümmten Stäben und/oder bei verteilten Folgelasten treten jedoch auch diejenigen Integrale auf, bei denen die Summe der Ableitungen eins oder drei ist. Diese Integrale werden für vier Lagerungen in [5] angegeben. Sind die Koeffizienten in den Bewegungsgleichungen des Systems selbst ortsabhängig, so lassen sich die auftretenden Integrale i. a. nur noch auf numerischem Wege berechnen. Dies wäre zwar auch für die erstgenannten Integrale möglich, doch führt die Verwendung geschlossener Lösungen zu einer Erhöhung der Genauigkeit bzw. Verringerung der Rechenzeit.

In der vorliegenden Arbeit werden die Integrale über je zwei Balkenfunktionen bzw. deren Ableitungen bei beliebiger Ableitungssumme angegeben. Die Integrale werden durch die nullte bis dritte Ableitung der Balkenfunktionen dargestellt, so daß die Ergebnisse für jeden beliebigen Lagerungsfall gelten. Es werden außerdem Formeln zur Berechnung obiger Integrale für den Fall angegeben, daß der Integrand zusätzlich den Faktor x^n mit $n \geqq 0$, ganz enthält; diese Integrale treten auf, wenn die Koeffizienten der Bewegungsgleichungen durch Polynome der Ortskoordinate darstellbar sind.

2. Das Integral erster Ordnung

Die k-te Balkenfunktion $y_k = y_k(\lambda_k x)$ mit $x \in [a, a + 1]$, $a \in R$ genügt als k-te Eigenschwingungsform des Biegebalkens konstanter Steifigkeit und Massenbelegung der Differentialgleichung

$$y_k^{(4)} - \lambda_k^4 y_k = 0 \tag{1}$$

bei gleichzeitiger Erfüllung von je zwei linearen, homogenen Randbedingungen bei $x = a$ und $x = a + 1$, durch welche die abzählbar unendlich vielen Eigenwerte λ_k festgelegt werden. Es sind y_k, λ_k, x, a einheitenfreie Größen; Ableitungen nach x werden durch hochgestellte, in Klammern gesetzte, positive Indizes gekennzeichnet; die Größe a ist beliebig wählbar. Für die Differentiation nach dem Eigenwert λ_k gilt die Regel

$$\frac{\partial}{\partial \lambda_k} [y_k^{(\sigma)}(\lambda_k x)] = \frac{1}{\lambda_k} [\sigma y_k^{(\sigma)}(\lambda_k x) + x y_k^{(\sigma+1)}(\lambda_k x)] . \tag{2}$$

Viermalige partielle Integration von $y_r^{(p)} \cdot y_s^{(q)}$ führt mit (1) auf das Ausgangsintegral zurück und man erhält für $r \neq s$ die Darstellung

$$\int y_r^{(p)} y_s^{(q)} \, dx = \frac{1}{\lambda_r^4 - \lambda_s^4} [y_r^{(p+3)} y_s^{(q)} - y_r^{(p+2)} y_s^{(q+1)} + y_r^{(p+1)} y_s^{(q+2)} - y_r^{(p)} y_s^{(q+3)}] . \tag{3}$$

Das Integral über zwei Balkenfunktionen bzw. deren Ableitungen existiert auch für $r = s$, wobei hierfür in (3) nicht nur der Nenner, sondern auch der Zähler verschwindet. Durch Differentiation nach λ_r, sowohl des Zählers unter Beachtung von (2), als auch des Nenners und anschließendem Grenzübergang λ_r gegen λ_s folgt für $r = s$ die Darstellung

$$\begin{aligned}\int y_s^{(p)} y_s^{(q)} \, dx = {} & \frac{1}{4\lambda_s^4} \{[(p + 3)\, y_s^{(p+3)} y_s^{(q)} - (p + 2)\, y_s^{(p+2)} y_s^{(q+1)} + \\ & + (p + 1)\, y_s^{(p+1)} y_s^{(q+2)} - p y_s^{(p)} y_s^{(q+3)}] + \\ & + x\, [y_s^{(p+4)} y_s^{(q)} - y_s^{(p+3)} y_s^{(q+1)} + y_s^{(p+2)} y_s^{(q+2)} - y_s^{(p+1)} y_s^{(q+3)}]\} .\end{aligned} \tag{4}$$

Von der Gültigkeit der Beziehungen (3) und (4) überzeugt man sich unmittelbar durch Differentiation nach x. Im Anhang A sind die Integrale (3) und (4) für $p, q = 0, 1, 2, 3$ angegeben.

3. Das Integral m-ter Ordnung

Das Integral m-ter Ordnung der Funktion

$$J_{rs}^{p,q} \stackrel{\text{def}}{=} y_r^{(p)} \cdot y_s^{(q)} , \qquad p, q = 0, 1, 2, 3 \tag{5}$$

wird definiert als

$$^{(-m)}J_{rs}^{p,q} \stackrel{\text{def}}{=} \underbrace{\int \dots \int}_{m\text{-fach}} J_{rs}^{p,q} (dx)^m , \qquad m = 0, 1, 2, \dots . \tag{6}$$

In der Schreibweise (5), (6) läßt sich (3) mit $N_{rs} = \lambda_r^4 - \lambda_s^4$ als Integral erster Ordnung über $J_{rs}^{p,q}$ darstellen als

$$^{(-1)}J_{rs}^{p,q} = [J_{rs}^{p+3,q} - J_{rs}^{p+2,q+1} + J_{rs}^{p+1,q+2} - J_{rs}^{p,q+3}]/N_{rs} . \tag{7}$$

Da sich aber $^{(-2)}J_{rs}^{p,q}$ durch Integration von $^{(-1)}J_{rs}^{p,q}$ ergibt, und da jeder Summand in (7) wieder vom gleichen Typ wie $J_{rs}^{p,q}$ selbst ist, kann (7) als Integrationsformel verwendet werden, um die Integrale zweiter und höherer Ordnung über $J_{rs}^{p,q}$ zu erhalten. Für beliebiges $m \geqq 0$, ganz und $r \neq s$ erhält man mit (5) die Darstellung

$$^{(-m)}J_{rs}^{p,q} = \frac{1}{N_{rs}^m} \sum_{t=0}^{3m} (-1)^t \, c_t^{(m)} y_r^{(p+3m-t)} y_s^{(q+t)} \tag{8}$$

mit den Koeffizienten

$$\begin{aligned} c_0^{(m)} &= c_0^{(m-1)} , \\ c_1^{(m)} &= c_1^{(m-1)} + c_0^{(m-1)} , \\ c_2^{(m)} &= c_2^{(m-1)} + c_1^{(m-1)} + c_0^{(m-1)} , \\ c_t^{(m)} &= \sum_{j=0}^{3} c_{t-j}^{(m-1)} \quad \text{für} \quad 3 \leqq t \leqq 3m - 3 , \\ c_{3m-2}^{(m)} &= c_{3m-3}^{(m-1)} + c_{3m-4}^{(m-1)} + c_{3m-5}^{(m-1)} , \\ c_{3m-1}^{(m)} &= c_{3m-3}^{(m-1)} + c_{3m-4}^{(m-1)} , \\ c_{3m}^{(m)} &= c_{3m-3}^{(m-1)} . \end{aligned} \tag{9}$$

Das Integral (8) ist bezüglich p und q symmetrisch, da für die Koeffizienten $c_{3m}^{(m)} = c_0^{(m)}$, $c_{3m-1}^{(m)} = c_1^{(m)}$ usw. gilt. Die Koeffizienten (9) sind für $m = 1$ bis $m = 4$ im Anhang B angegeben; mit ihnen lassen sich aus (8) die entsprechenden Integrale m-ter Ordnung für $r \neq s$ unmittelbar angeben. Der Beweis von (8), (9) erfolgt durch vollständige Induktion: Gln. (9), (10) gelten für $m = 1$ und $m = 2$; das beliebige m wird durch $m + 1$ ersetzt, der erhaltene Ausdruck nach x differenziert und gezeigt, daß das Ergebnis mit $^{(-m)}J_{rs}^{p,q}$ identisch ist.

Für $r = s$ verschwinden in (8) Zähler und Nenner und es liegt ein unbestimmter Ausdruck der Form $0^m/0^m$ vor. Durch m-fache Differentiation von Zähler und Nenner nach λ_r und anschließenden Grenzübergang λ_r gegen λ_s folgt aus (8) das Integral für $r = s$. Für die m-fache Differentiation des Zählers nach λ_k gilt (10) als Verallgemeinerung von (2):

$$\frac{\partial^m}{\partial \lambda_k^m} [y_k^{(\sigma)}(\lambda_k x)] = \frac{1}{\lambda_k^m} \sum_{j=0}^{m} \binom{m}{j} \frac{\sigma!}{(\sigma - m + j)!} x^j y_k^{(\sigma+j)} . \tag{10}$$

Der Beweis von (10) erfolgt durch vollständige Induktion: Gl. (10) gilt für $m = 1$, vergl. (2); das beliebige m wird durch $m + 1$ ersetzt und es wird gezeigt, daß der erhaltene Ausdruck identisch ist mit dem, der durch Anwendung von (2) auf (10) gewonnen wird. Die m-te Ableitung von N_{rs}^m nach λ_r ergibt sich für $r = s$ als

$$\left[\frac{\partial^m N_{rs}^m}{\partial \lambda_r^m}\right]_{r=s} = m!(4\lambda_s^3)^m . \tag{11}$$

Unter Verwendung von (10) und (11), wobei $\sigma = p + 3m - t$ zu setzen ist, und durch Vertauschen der Reihenfolge der Summationen erhält man

$$^{(-m)}J_{ss}^{p,q} = \frac{1}{m!(4\lambda_s^4)^m} \sum_{j=0}^{m} \left[\binom{m}{j} x^j \sum_{t=0}^{3m} (-1)^t \gamma_{t,j}^{(m,p)} y_s^{(p+3m+j-t)} y_s^{(q+t)}\right] \tag{12}$$

mit

$$\gamma_{t,j}^{(m,p)} = \frac{(p + 3m - t)!}{(p + 2m + j - t)!} c_t^{(m)} . \tag{13}$$

Im Gegensatz zu den Koeffizienten $c_t^{(m)}$, die nur von m abhängen, sind die Koeffizienten (13) Funktionen von m und p. Im Anhang C sind sie für $p = 0, 1, 2, 3$ und $m = 1$ bis $m = 4$ angegeben.

4. Der Faktor x^n im Integranden

Unter Verwendung der Schreibweise (5), (6) erhält man durch partielle Integration von $x^n y_r^{(p)} y_s^{(q)}$ mit $n \geqq 0$, ganz, die Rekursionsformel

$$\int J_{rs}^{p,q} x^n \, dx = {}^{(-1)}J_{rs}^{p,q} x^n - n \int {}^{(-1)}J_{rs}^{p,q} x^{n-1} \, dx , \tag{14}$$

die für beliebiges r, s Gültigkeit besitzt. Wiederholte Anwendung von (14) führt mit (6) auf die Darstellung

$$\int x^n y_r^{(p)} y_s^{(q)} \, dx = n! \sum_{k=0}^{n} (-1)^k \frac{x^{n-k}}{(n - k)!} \cdot {}^{(-k-1)}J_{rs}^{p,q} , \tag{15}$$

von deren Gültigkeit man sich durch Differentiation nach x unmittelbar überzeugt. Für $r \neq s$ sind die Integrale $(k + 1)$-ter Ordnung durch (8), (9) gegeben und in (15) tritt in jedem Summanden ein Faktor x^{n-k} auf. Für $r = s$ sind die Integrale $(k + 1)$-ter Ordnung durch (12), (13) gegeben, so daß in jedem Summanden außer dem Faktor x^{n-k} zusätzliche Potenzen von x entsprechend (12) auftreten und eine Ordnung nach Potenzen von x möglich wäre. Dies führt jedoch für beliebiges n, p, q zu sehr unübersichtlichen Formeln, so daß es vorteilhafter erscheint, die $^{(-k-1)}J_{rs}^{p,q}$ nach (12), (13) für den in der Anwendung benötigten Fall zu berechnen, in (15) einzusetzen und — falls erforderlich — nach Potenzen von x zu ordnen.

5. Zusammenfassung

Die für vorgegebenes p, q in den unbestimmten Integralen (3), (4) und (8), (12) auftretenden Ableitungen vierter und höherer Ordnung lassen sich mit (1) stets durch die Ableitungen nullter bis dritter Ordnung ausdrücken. Damit sind aber auch die bestimmten Integrale stets durch die für den konkreten Lagerungsfall bekannten, nur von den Eigenwerten abhängigen Ableitungen nullter bis dritter Ordnung an den Rändern $x = a$ und $x = a + 1$ darstellbar. Im Gegensatz zu den bisher in der Literatur angegebenen Integralen gelten die hier erhaltenen Ergebnisse somit für jeden beliebigen Lagerungsfall, einschließlich solcher Fälle, daß an den Trägerrändern diskrete Federn, Massen und Dämpfungselemente vorhanden sind.

Anhang A: Die Integrale erster Ordnung

Der obere Ausdruck gilt jeweils für $r \neq s$, der untere für $r = s$; Striche kennzeichnen Ableitungen nach x und es bedeuten $N_{rs} = \lambda_r^4 - \lambda_s^4$ $N_{ss} = 4\lambda_s^4$.

$$\int y_r y_s \, dx = \begin{cases} [y_r''' y_s - y_r'' y_s' + y_r' y_s'' - y_r y_s''']/N_{rs} \\ [3y_s''' y_s - y_s'' y_s' + x\{\lambda_s^4 y_s^2 - 2y_s''' y_s' + y_s''^2\}/N_{ss} \end{cases}$$

$$\int y_r y_s' \, dx = \begin{cases} [y_r''' y_s' - y_r'' y_s'' + y_r' y_s''' - \lambda_s^4 y_r y_s]/N_{rs} \\ [4y_s''' y_s' - 2y_s''^2]/N_{ss} \end{cases}$$

$$\int y_r y_s'' \, dx = \begin{cases} [y_r''' y_s'' - y_r'' y_s''' + \lambda_s^4(y_r' y_s - y_r y_s')]/N_{rs} \\ [y_s''' y_s'' + \lambda_s^4 y_s' y_s + x\{\lambda_s^4(2y_s'' y_s - y_s'^2) - y_s'''^2\}]/N_{ss} \end{cases}$$

$$\int y_r y_s''' \, dx = \begin{cases} [y_r''' y_s''' - \lambda_s^4(y_r'' y_s - y_r' y_s' + y_r y_s'')]/N_{rs} \\ [3y_s'''^2 - \lambda_s^4(2y_s'' y_s - y_s'^2)]/N_{ss} \end{cases}$$

$$\int y_r' y_s \, dx = \begin{cases} [\lambda_r^4 y_r y_s - y_r''' y_s' + y_r'' y_s'' - y_r' y_s''']/N_{rs} \\ [4y_s''' y_s' - 2y_s''^2]/N_{ss} \end{cases}$$

$$\int y_r' y_s' \, dx = \begin{cases} [\lambda_r^4 y_r y_s' - y_r''' y_s'' + y_r'' y_s''' - \lambda_s^4 y_r' y_s]/N_{rs} \\ [-y_s''' y_s'' + 3\lambda_s^4 y_s' y_s + x\{\lambda_s^4(y_s'^2 - 2y_s'' y_s) + y_s'''^2\}]/N_{ss} \end{cases}$$

$$\int y_r' y_s'' \, dx = \begin{cases} [\lambda_r^4 y_r y_s'' - y_r''' y_s''' + \lambda_s^4(y_r'' y_s - y_r' y_s')]/N_{rs} \\ [-3y_s'''^2 + \lambda_s^4(6y_s'' y_s - y_s'^2)]/N_{ss} \end{cases}$$

$$\int y_r' y_s''' \, dx = \begin{cases} [\lambda_r^4 y_r y_s''' - \lambda_s^4(y_r''' y_s - y_r'' y_s' + y_r' y_s'')]/N_{rs} \\ [y_s''' y_s + y_s'' y_s' + x\{-\lambda_s^4 y_s^2 + 2y_s''' y_s' - y_s''^2\}]/N_{ss} \end{cases}$$

$$\int y_r'' y_s \, dx = \begin{cases} [\lambda_r^4(y_r'' y_s - y_r' y_s' + y_r y_s'') - y_r''' y_s''']/N_{rs} \\ [y_s''' y_s'' + \lambda_s^4 y_s' y_s + x\{\lambda_s^4(2y_s'' y_s - y_s'^2) - y_s'''^2\}]/N_{ss} \end{cases}$$

$$\int y_r'' y_s' \, dx = \begin{cases} [\lambda_r^4(y_r' y_s' - y_r y_s'') + y_r''' y_s''' - \lambda_s^4 y_r'' y_s]/N_{rs} \\ [-3y_s'''^2 + \lambda_s^4(6y_s'' y_s - y_s'^2)]/N_{ss} \end{cases}$$

$$\int y_r'' y_s'' \, dx = \begin{cases} [\lambda_r^4(y_r' y_s'' - y_r y_s''') + \lambda_s^4(y_r''' y_s - y_r'' y_s')]/N_{rs} \\ [3y_s'' y_s' - y_s''' y_s + x\{\lambda_s^4 y_s^2 + y_s''^2 - 2y_s''' y_s'\}]/N_{ss} \end{cases}$$

$$\int y_r'' y_s''' \, dx = \begin{cases} [\lambda_r^4 y_r' y_s''' - \lambda_r^4 \lambda_s^4 y_r y_s + \lambda_s^4(y_r''' y_s' - y_r'' y_s'')]/N_{rs} \\ [8y_s''' y_s' - 2y_s''^2 - 4\lambda_s^4 y_s^2]/N_{ss} \end{cases}$$

$$\int y_r''' y_s \, dx = \begin{cases} [\lambda_r^4(y_r'' y_s - y_r' y_s' + y_r y_s'') - y_r''' y_s''']/N_{rs} \\ [3y_s'''^2 - \lambda_s^4(2y_s'' y_s - y_s'^2)]/N_{ss} \end{cases}$$

$$\int y_r''' y_s' \, dx = \begin{cases} [\lambda_r^4(y_r'' y_s' - y_r' y_s'' + y_r y_s''') - \lambda_s^4 y_r''' y_s]/N_{rs} \\ [y_s''' y_s + y_s'' y_s' + x\{-\lambda_s^4 y_s^2 + 2y_s''' y_s' - y_s''^2\}]/N_{ss} \end{cases}$$

$$\int y_r''' y_s'' \, dx = \begin{cases} [\lambda_r^4(y_r'' y_s'' - y_r' y_s''') + \lambda_r^4 \lambda_s^4 y_r y_s - \lambda_s^4 y_r''' y_s']/N_{rs} \\ [8y_s''' y_s' - 2y_s''^2 - 4\lambda_s^4 y_s^2]/N_{ss} \end{cases}$$

$$\int y_r''' y_s''' \, dx = \begin{cases} [\lambda_r^4 y_r'' y_s''' - \lambda_r^4 \lambda_s^4(y_r' y_s - y_r y_s') - \lambda_s^4 y_r''' y_s'']/N_{rs} \\ [3y_s''' y_s'' - \lambda_s^4 y_s' y_s + x\{-\lambda_s^4(2y_s'' y_s - y_s'^2) + y_s'''^2\}]/N_{ss} \end{cases}$$

Anhang B: Die Koeffizienten $c_t^{(m)}$

$m = 1$	$m = 2$	$m = 3$	$m = 4$
			1
			4
		1	10
	1	3	20
	2	6	31
1	3	10	40
1	4	12	44
1	3	12	40
1	2	10	31
	1	6	20
		3	10
		1	4
			1

Anhang C: Die Koeffizienten $\gamma_{t,j}^{(m,p)}$ für $p = 0$

m	t	j = 0	1	2	3	4
1	0	3	1			
	1	2	1			
	2	1	1			
	3	0	1			
2	0	30	6	1		
	1	40	10	2		
	2	36	12	3		
	3	24	12	4		
	4	6	6	3		
	5	0	2	2		
	6	0	0	1		
3	0	504	72	9	1	
	1	1008	168	24	3	
	2	1260	252	42	6	
	3	1200	300	60	10	
	4	720	240	60	12	
	5	288	144	48	12	
	6	60	60	30	10	
	7	0	12	12	6	
	8	0	0	3	3	
	9	0	0	0	1	
4	0	11880	1320	132	12	1
	1	31680	3960	440	44	4
	2	50400	7200	900	100	10
	3	60480	10080	1440	180	20
	4	52080	10416	1736	248	31
	5	33600	8400	1680	280	40
	6	15840	5280	1320	264	44
	7	4800	2400	800	200	40
	8	744	744	372	124	31
	9	0	120	120	60	20
	10	0	0	20	20	10
	11	0	0	0	4	4
	12	0	0	0	0	1

Die Koeffizienten $\gamma_{t,j}^{(m,p)}$ für $p = 1$

m	t	j = 0	1	2	3	4
1	0	4	1			
	1	3	1			
	2	2	1			
	3	1	1			
2	0	42	7	1		
	1	60	12	2		
	2	60	15	3		
	3	48	16	4		
	4	18	9	3		
	5	4	4	2		
	6	0	1	1		
3	0	720	90	10	1	
	1	1512	216	27	3	
	2	2016	336	48	6	
	3	2100	420	70	10	
	4	1440	360	72	12	
	5	720	240	60	12	
	6	240	120	40	10	
	7	36	36	18	6	
	8	0	6	6	3	
	9	0	0	1	1	
4	0	17160	1716	156	13	1
	1	47520	5280	528	48	4
	2	79200	9900	1100	110	10
	3	100800	14400	1800	200	20
	4	93744	15624	2232	279	31
	5	67200	13440	2240	320	40
	6	36960	9240	1848	308	44
	7	14400	4800	1200	240	40
	8	3720	1860	620	155	31
	9	480	480	240	80	20
	10	0	0	60	30	10
	11	0	0	8	8	4
	12	0	0	0	1	1

Die Koeffizienten $\gamma_{t,j}^{(p,m)}$ für $p = 2$

m	t	j				
		0	1	2	3	4
1	0	5	1			
	1	4	1			
	2	3	1			
	3	2	1			
2	0	56	8	1		
	1	84	14	2		
	2	90	18	3		
	3	80	20	4		
	4	36	12	3		
	5	12	6	2		
	6	2	2	1		
3	0	990	110	11	1	
	1	2160	270	30	3	
	2	3024	432	54	6	
	3	3360	560	80	10	
	4	2520	504	84	12	
	5	1440	360	72	12	
	6	600	200	50	10	
	7	144	72	24	6	
	8	18	18	9	3	
	9	0	2	2	1	
4	0	24024	2184	182	14	1
	1	68640	6864	624	52	4
	2	118800	13200	1320	120	10
	3	158400	19800	2200	220	20
	4	156240	22320	2790	310	31
	5	120960	20160	2880	360	40
	6	73920	14784	2464	352	44
	7	33600	8400	1680	280	40
	8	11160	3720	930	186	31
	9	2400	1200	400	100	20
	10	240	240	120	40	10
	11	0	24	24	12	4
	12	0	0	2	2	1

Die Koeffizienten $\gamma_{t,j}^{(m,p)}$ für $p = 3$

m	t	j				
		0	1	2	3	4
1	0	6	1			
	1	5	1			
	2	4	1			
	3	3	1			
2	0	72	9	1		
	1	112	16	2		
	2	126	21	3		
	3	120	24	4		
	4	60	15	3		
	5	24	8	2		
	6	6	3	1		
3	0	1320	132	12	1	
	1	2970	330	33	3	
	2	4320	540	60	6	
	3	5040	720	90	10	
	4	4032	672	96	12	
	5	2520	504	84	12	
	6	1200	300	60	10	
	7	360	120	30	6	
	8	72	36	12	3	
	9	6	6	3	1	
4	0	32760	2730	210	15	1
	1	96096	8736	728	56	4
	2	171600	17160	1560	130	10
	3	237600	26400	2640	240	20
	4	245520	30690	3410	341	31
	5	201600	28800	3600	400	40
	6	133056	22176	3168	396	44
	7	67200	13440	2240	320	40
	8	26040	6510	1302	217	31
	9	7200	2400	600	120	20
	10	1200	600	200	50	10
	11	96	96	48	16	4
	12	0	6	6	3	1

Literatur

1 SOROKIN, E. S., Dinamika mežduetažnych perekrytyj, Moskva 1941, S. 218 bis 227.
2 BISHOP, R. E. D., JOHNSON, D. C., The mechanics of vibration, Cambridge 1960, Kap. 7.
3 VLASSOV, B. Z., Tonkostennye prostranstvennye sistemy, Moskva 1958, S. 98–102.
4 FILIPPOV, A. P., Kolebanija deformiruemych sistem, Moskva 1970, S. 78–82.
5 STEPANOV, R. D., O nekotorych kvadraturach baločnych fundamental'nych funkcij, Inženernyj sbornik **26** (1958), S. 94–100.

Eingereicht am 9. 6. 1980

Anschrift: Dr.-Ing. OSWALD BECKER, Ingenieurhochschule Zittau, Abt. Mathematik, 88 Zittau, Theodor-Körner-Allee 16, DDR

ZAMM **61**, 272–273 (1981)

M. RIEDEL

On Bosch's Characterization of the Exponential Distribution Function

Let F be a distribution function of a non-negative random variable which is not concentrated at zero. For $t > 0$ we define a family p_t of discrete probability distributions concentrated on all non-negative integers k by

$$p_t(k) = F((k + 1)\, t) - F(kt)\,.$$

If F is the exponential distribution function with parameter $\lambda > 0$ then p_t is the geometric probability distribution with parameter $1 - p_t(0) = \exp(-\lambda t)$ as can easily be seen.

A characterization of the exponential distribution function by this property was obtained by K. BOSCH [1]; he proved the following result.

Theorem 1: *F is the exponential distribution function with an arbitrary parameter $\lambda > 0$ if and only if p_t is a geometric probability distribution for all $t > 0$.*

Remark: From BOSCH's proof it follows that the theorem remains true if t is restricted to $t = 1/n$ for all natural numbers n.

We now intend to generalize the theorem in another direction.

Theorem 2: *Let $l > 1$ be some natural number. F is the exponential distribution function if and only if for all $t > 0$, $p_t(k)$ coincides with the corresponding jumps of a geometric probability distribution for $k = 0, 1, l$.*

Remark:

1. For $l = 2$ theorem 2 follows from theorem 2.1.2 of [4].
2. The author is studying whether it is possible to replace 1 by an arbitrary number $m \neq l$.

For the proof we need the following lemma (see [2]), whose proof we give for the sake of completeness. Obviously, we have

$$\log_2 t = \frac{\log t}{\log 2}$$

where the natural logarithm of t is denoted by $\log t$.

Lemma: *Let $g\colon (0, \infty) \to (0, 1)$ be a function which satisfies*

$$g(2t) = g^2(t)\,. \tag{1}$$

Then g is of the form

$$g(t) = \exp\left(-t \cdot 2^{C(\log_2 t)}\right) \tag{2}$$

where C is a periodic function with period 1.

Proof: Set

$$A(t) := \log(-\log g(t))\,.$$

Then (1) can be written as

$$A(2t) = A(t) + \log 2\,.$$

Using the notations $s := \log_2 t$ and $B(s) := A(2^s)/\log 2$ we obtain

$$B(s+1) = B(s) + 1 .$$

This is a difference equation with continuous argument and its general solution is given by

$$B(s) = s + C(s)$$

where C is an arbitrary periodic function with period 1. Hence, for g we get the representation (2).

Proof of theorem 2: We consider only the sufficiency because the necessity is trivial. By assumption we have

$$p_t(1) = p_t(0)(1 - p_t(0)) \tag{3}$$

and

$$p_t(l) = p_t(0)(1 - p_t(0))^{l-1} . \tag{4}$$

Using the definition of p_t we see that (3) and (4) are equivalent to

$$F(2t) - F(t) = F(t)(1 - F(t)) \tag{3'}$$

and

$$F(lt) - F((l-1)\,t) = F(t)(1 - F(t))^{l-1} , \tag{4'}$$

respectively. We write $G(t) := 1 - F(t)$ and get from (3′)

$$G(2t) = G^2(t) . \tag{3''}$$

Note that $0 < G(t) < 1$ is satisfied: Assuming $G(t_0) = 0$ ($G(t_0) = 1$) for some $t_0 > 0$ we obtain from (3″) $G(2t_0) = 0$, $G(t_0/2) = 0$ ($G(2t_0) = 1$, $G(t_0/2) = 1$). Then $G \equiv 0$ ($G \equiv 1$) because G is monotonously decreasing. But $G \equiv 0$ is impossible since $G(0) = 1 - F(0) = 1$. On the other hand $G \equiv 1$ contradicts the assumption that F is non-degenerate.

Now we can apply the lemma to $g := G$ and get for F the representation

$$F(t) = 1 - \exp\left(-t \cdot 2^{C(\log_2 t)}\right) \tag{5}$$

with a periodic function C with period 1. Using the representation (5) we can write (4′) in the form

$$\exp\left(-(l-1)\,t \cdot 2^{C(\log_2 t + a)}\right) - \exp\left(-lt \cdot 2^{C(\log_2 t + b)}\right) =$$
$$= \exp\left(-(l-1)\,t \cdot 2^{C(\log_2 t)}\right)\left(1 - \exp\left(-t \cdot 2^{C(\log_2 t)}\right)\right) \tag{6}$$

where

$$a := \log_2 (l-1) \quad \text{and} \quad b := \log_2 l .$$

In order to investigate the function C we use the TAYLOR expansion to first order. For a natural number n and for real numbers t_0, d with $t_0 > 0$ we have

$$C(\log_2 2^{-n} t_0 + d) = C(\log_2 t_0 + d) .$$

Putting $t = t_0 2^{-n}$ in (6), multiplying by 2^n and tending $n \to \infty$ we get

$$l \cdot 2^{C(\log_2 t_0 + b)} - (l-1)\, 2^{C(\log_2 t_0 + a)} = 2^{C(\log_2 t_0)} . \tag{7}$$

Substituting $C(\log_2 t_0 + b)$ from (7) into (6) we obtain by simple computations

$$C(\log_2 t_0 + a) = C(\log_2 t_0) .$$

Hence, the function C has the periods 1 and a. Now we have to distinguish two cases concerning $a = \log_2 (l-1)$.

a) a is irrational. In this case C has periods of the form $m + na$ where m, n are integers. Since these periods are dense on $(-\infty, \infty)$ (cf. e.g. [3] lemma 4) we can represent F (with a certain constant $\lambda > 0$) by

$$F(t) = 1 - \exp(-\lambda t)$$

where t belongs to a dense set on $(0, \infty)$. Since F is a distribution function it follows that F is the exponential distribution function.

b) a is rational, i.e. $l - 1 = 2^r$ for some natural number r and $a = r$. Then (6) is equivalent to

$$C(\log_2 t + b) = C(\log_2 t) .$$

Since b is irrational we conclude the statement as in case a).

References

1 BOSCH, K., Eine Charakterisierung der Exponentialverteilung, ZAMM 57 (1977), 609–610.
2 DALLAS, A., On the minimum of a random sample, Math. Operationsforsch. Statist., Ser. Statistics 8 (1977), 511–513.
3 EATON, M. L., Characterization of distributions by the identical distribution of linear forms, J. Appl. Probab. 3 (1966), 481–494
4 GALAMBOS, J.; SAMUEL, K., Characterization of Probability Distributions, Springer-Verlag, Berlin-Heidelberg-New York 1978, 24.

Eingereicht: 7. 5. 1980, korrigierte Fassung 14. 11. 1980

Anschrift: Dr. M. RIEDEL, Karl-Marx-Universität Leipzig Sektion Mathematik, Karl-Marx-Platz 10–12, DDR-7010 Leipzig

BUCHBESPRECHUNGEN

Lommatsch, K. (Hrsg.), Anwendungen der linearen parametrischen Optimierung. Berlin, Akademie-Verlag 1979. X, 189 S., M 38,–.

Lineare Optimierung wird seit Jahrzehnten in außerordentlich großem Maße in den meisten einer mathematischen Erfassung zugänglichen Zweigen der Natur- und Wirtschaftswissenschaften verwendet und verdankt ihren großen Einsatz nicht zuletzt der Verbreitung elektronischer Rechenanlagen. Dementsprechend ist auch die mathematische Literatur über Optimierungsaufgaben angewachsen, aber eine wichtige und naheliegende Fragestellung hat sich der mathematischen Behandlung gegenüber bisher als recht spröde erwiesen, nämlich die Frage nach dem Einfluß von Fehlern und Ungenauigkeiten in den zur Verfügung gestellten Daten. Dabei zeigt sich ein grundlegender Unterschied, ob sich die Ungenauigkeiten auf die Koeffizienten der Zielfunktion oder der „rechten Seiten" der Nebenbedingungen beziehen (das ist der einfachere Fall der Optimierung mit „fixierter Matrix") oder auf die „Koeffizienten-Matrix" (das ist der schwierigere Fall). Diese Fragen gehören zur „Sensitivitätsanalyse", die hier in den allgemeineren Rahmen der „parametrischen Optimierung" gestellt wird. Dabei soll das Wort „parametrische Optimierung" nicht bedeuten, daß einzelne Parameter und irgendwelche Elemente optimal bestimmt werden sollen, sondern die parametrische Optimierung beschäftigt sich mit der Untersuchung und Kennzeichnung bestimmter Teilmengen im Parameterraum, wie z. B. der zulässigen Parametermenge, der Lösbarkeitsmenge u. a. Zu den Betrachtungen, die in diesen Rahmen fallen, gehört unter anderem das Studium der Abhängigkeit der Lösungen von den Ausgangsdaten und damit die Stabilität von Lösungen, insbesondere die lokale Stabilität, der Einfluß nicht gesicherter Ausgangsdaten, Vergleich verschiedener Optimierungsaufgaben und dergleichen.

Das von LOMMATSCH herausgegebene Buch enthält 10 Kapitel, die von den Herrn LOMMATSCH, KLATTE, GUDDAT, TAMMER, DAMM, KOPIELSKI, BANK, MANDEL, NOWACK, MELZER, NOŽIČKA, WENDLER verfaßt wurden. Das erste von Lommatsch geschriebene Kapitel dient einem allgemeinen Überblick und einer Bereitstellung von Begriffen und Ergebnissen der linearen parametrischen Optimierung unter Bezugnahme auf den Ideenkreis von NOŽIČKA, dem auch das ganze Buch gewidmet ist. Wegen der Beweise wird verschiedentlich auf das Buch von NOŽIČKA „Theorie der linearen parametrischen Optimierung" verwiesen.

Das Buch zeigt die Durchführung der Ideen von NOŽIČKA in einer Reihe von Arbeiten seiner Schüler; Kapitel 2 befaßt sich mit dem Fall von Variationen der Koeffizientenmatrix, für den noch keine geschlossene Theorie vorliegt; es werden 2 verschiedene Wege beschritten, wobei man im Fall von nur einer vollparametrisierten Zeile weitere Aussagen machen kann. Kapitel 3 und 4 behandeln Vektor- und stochastische Optimierung. Prinzipiell ist es möglich, eine in gewissem Sinne maximale Menge aller als „optimal" zu bezeichnenden Lösungen aufzustellen, was aber praktisch einen sehr großen Aufwand verursachen kann. Kapitel 5 geht auf optimale Steuerungen, Bang-bang-Prinzip und obere Schranken für die Anzahl der Umkehrpunkte ein und Kapitel 6 auf Aufteilungsverfahren zur Lösung spezieller Typen von Optimierungsaufgaben, bei denen die Zielfunktion bezüglich zweier Variablengruppen separabel ist, speziell auf gemischt ganzzahlig-lineare Optimierung und gewisse Klassen indefiniter quadratischer Optimierungen. Bei quadratischer Zielfunktion in Kapitel 7 wird unter anderem die Lage der Optimalpunkte und in Kapitel 8 einige Eigenschaften registrierter Bimatrixspiele, sog. Lin-Opt-Spiele behandelt. Kapitel 9 befaßt sich mit dem STEINER-WEBER-Problem zur Bestimmung eines optimalen Standortes mit minimalen Transportkosten im n-dimensionalen Raum R^n und Beschreibung von Lösungsverfahren. Kapitel 10 entwickelt ein lineares parametrisches Modell für Gemüseernte, die teilweise von Hand und von einem optimal zu wählenden Zeitpunkt an mit einer Erntemaschine vorgenommen werden soll. Es werden 2 einfache zweiparametrige Beispiele gegeben, aber ohne Auswertung der Ergebnisse. Es erscheint fraglich, ob diese Beispiele geeignet sind, den Praktiker von dem Nutzen der Theorie zu überzeugen.

Das Buch gibt einen ausgezeichneten theoretischen Überblick über den gegenwärtigen Stand der Untersuchung linearer parametrischer Optimierungen und kann sehr empfohlen werden. Nur müßte der Titel statt „Anwendungen der linearen parametrischen Optimierung" richtig heißen: „Theorie der linearen parametrischen Optimierung mit Anwendungen".

Hamburg L. COLLATZ

Babič, V. M. / Kirpičnikova, N. Y., The Boundary-Layer Method in Diffraction Problems. Springer Series in Electrophysics 3. Translation from the Russian. Berlin-Heidelberg-New York, Springer-Verlag 1979. VI, 140 S., DM 39,50. US $ 21.80. ISBN 3-540-09605-1

In diesem Buch werden Methoden zur Behandlung singulär gestörter Differentialgleichungsprobleme angewendet auf die Lösung von Problemen der skalaren Wellengleichung mit ortsabhängiger Geschwindigkeit; Probleme dieser Art treten bei der Behandlung von Beugungserscheinungen auf. Im Gültigkeitsbereich der Strahlenoptik existiert die reguläre Entwicklung des Strahlenansatzes mit der inversen Kreisfrequenz als Störparameter. Diese Entwicklung versagt in der Nachbarschaft gewisser Mengen. Eine in diesen Grenzschichten konvergente Entwicklung nach dem Störparameter wird durch Einführung geeigneter gestreckter Koordinaten erreicht. Die Anpassung der Lösung in der Grenzschicht an die durch den Strahlenansatz dargestellte Lösung sorgt dann für die eindeutige Lösung des behandelten Problems. Es wird zunächst das Problem der Kaustik betrachtet. Hier dient die Kaustik als Grenzschicht. Durch Anpassung der Lösung in der Umgebung der Kaustik an den Strahlenansatz gewinnt man die Lösung der Aufgabe, bei gegebenen Koeffizienten des Strahlenansatzes der einfallenden Welle die Koffezienten im Strahlenansatz der ausstrahlenden Welle zu bestimmen. Als nächstes werden Oberflächenwellen betrachtet, was auf die Lösung der Wellengleichung mit der Randbedingung Null hinauskommt. Es ergeben sich die Erscheinungen der „Flüstergalleriewellen" und der „kriechenden Wellen". Der Zusammenhang mit den Untersuchungen von FRIEDLÄNDER-KELLER wird beleuchtet. Weiter werden Schwingungserscheinungen von Hohlräumen betrachtet, die in der Nähe eines Strahls konzentriert sind. Ausführlich werden die Erscheinungen der Beugung an einem glatten konvexen Körper behandelt, so das Entstehen von kriechenden Wellen in der Schattenzone un die FRESNELschen Erscheinungen in der Nähe der Begrenzungsstrahlen. Zum Schluß wird das Problem einer schwingenden Punktquelle in einem Medium variabler Geschwindigkeit betrachtet. Die Methode der Anpassung der in einer Umgebung des Quellpunktes existierenden Darstellung der Lösung an die Strahlendarstellung liefert die Koeffizienten des Strahlungsansatzes.

An Vorkenntnissen zum Verständnis sind für dieses Buch erforderlich: Grundkenntnisse der Analysis und der Funktionentheorie.

Das Buch ist zu empfehlen angewandten Mathematikern und mathematischen Physikern, besonders allen, die sich mit der theoretischen Behandlung von Beugungserscheinungen optischer und akustischer Wellen befassen.

Berlin E. HEYN

Brousse, P. (ed.), Structural Optimization. International Centre for Mechanical Sciences, CISM Courses and Lectures No. 237. Wien-New York, Springer-Verlag 1975. VIII, 210 S., 25 Abb., ÖS 234,—, DM 34,—. ISBN 3-211-81376-4

The purpose of this book is to show how mathematical programming techniques, in particular nonlinear (convex) programming, duality theory and Lagrange techniques can be introduced and applied to structural mechanics. It consists of three parts by three different authors. Nevertheless it pretents to be regarded as a connected textbook. In the referee's opinion the authors did not attain this intention completely in all parts. The impression the book makes is somewhat inhomogeneous. If the reader, however, is willing not to take much notice of this little disadvantage he may learn a lot of things concerning application of mathematical programming to technical problems. Observing that this aim represents an increasing tendency in mathematical programming application in the last tens of years it can be stated that book fills a gap in mathematical programming literature. Moreover, in the reviewer's mind technical, in particular mechanical subjects yielding objective functions and only few constraints are more suitable for calculations and form quite a natural field of application for mathematical programming, possibly even more than economics does. To show this may be the main value the book. — It starts with an introductory part "General Methods of Optimization in Structural Mechanics" by P. BROUSSE, where the basics of linear and nonlinear programming are given together with some simple applications (vibrating structures, elastic beam) and some outline of numerical methods. The second part "Optimization Theory in the Design of Elastic-Plastic Structures" is by A. CYRAS and treats elastic-plastic bodies under monotonically increasing and cyclic loading as well giving static and kinematic formulations in each case and the corresponding generalized Lagrange problems. The final part "Theory of Optimal Plastic Design of Structures" by M. A. SAVE at first deals with optimality conditions and their application, a conception which is successively generalized to movable loads, stepwise varying design unknowns, action of bodyforces, convex cost function and terminates in a general criterion for optimal structural design.

Berlin H. WEINERT

Phillips, O. M., The Dynamics of the Upper Ocean. 2nd edition. Cambridge, Cambridge University Press 1980. VII, 336 S., £ 7.95 A P/B. ISBN 0-521-29801-6

For the first edition, see ZAMM 59, 331 (1979). In the second edition, according to measurement data and theoretical developments of more then ten years, revised and broadened insight is given into processes in the ocean like wave interaction, generation of surface waves by the wind, internal waves, and turbulence.

The value of the book lies in the profound presentation of theory which remains all the time strongly directed to the real subject and which is supported and illustrated by a great number of measuremental results.

Berlin G. STOYAN

Esslinger, M. (Hrsg.), Schalenbeultagung Darmstadt. Vorträge und Diskussionsbeiträge. Braunschweig, DFVLR 1979. VI, 250 S., DM 45,—. Bezug durch: Herrn G. Löffler, Inst. für Strukturmechanik, DFVLR, PF 3267, D-3300 Braunschweig.

Ebenso wie im Tagungsheft der Merseburger Schalenbeultagung im Jahre 1976 werden auch im vorliegenden Band sowohl Praxisprobleme als auch neuere Forschungsergebnisse dargelegt. Die Beulprobleme aus der Baupraxis stehen mit aufgetretenen Schadensfällen an Silos und Großbehältern in Verbindung; Aufgaben des Flugzeugbaues werden nur in einem

Beiträge angesprochen. Neben der Berechnung kritischer Lasten werden in einigen Vorträgen auch konstruktive Maßnahmen erörtert. Experimentelle Ergebnisse werden für Kegelschalen und lange Kreiszylinderschalen mitgeteilt. Es ist sehr zu begrüßen, daß die Vorträge der Schalenbeultagungen von der Deutschen Forschungs- und Versuchsanstalt für Luft- und Raumfahrt allen Interessenten durch Veröffentlichung zugänglich gemacht werden.

Berlin L. Knöfel

Wang, C.-C., Mathematical Principles of Mechanics and Electromagnetism. Part A: Analytical and Continuum Mechanics. Part B: Electromagnetism and Gravitation. Mathematical Concepts and Methods in Science and Engg. 16, 17. New York-London, Plenum Press 1979. 424 S., ISBN 0-306-40211-4. ISBN 0-306-40212-2

Chao Cheng Wang ist ein jüngerer, aber hervorragender Vertreter jener Gruppe von Kontinuumsmechanikern, die ihre Wissenschaft in solchen abstrakten Fernen moderner Mathematik angesiedelt haben, daß man fragen mußte, wie sie sich denn nun die Lehre vorstellen. Auf diese Frage gibt das vorliegende Werk Antwort, denn es möchte auch Studenten höherer Semester zugängig sein. Es beschäftigt sich freilich nicht nur mit der Mechanik der Kontinua, sondern außerdem mit der von Systemen endlich vieler Freiheitsgrade sowie mit dem Elektromagnetismus und der allgemeinrelativistischen Gravitationstheorie; und zwar werden die fundamentalen Prinzipien dieser Disziplinen entwickelt und durch die Erörterung einiger ausgewählter Themen illustriert. Dabei verzichtet Verf. in keiner Weise auf begriffliche Klarheit und mathematische Strenge, wohl aber auf extreme mathematische Modernität. Damit wird im gewissen Sinne deutlich, was bei den eingangs erwähnten Aktivitäten jeweils auf das Konto des einen oder des anderen kommt. Außerdem zeigt sich, daß eine stärkere Hinwendung zur konventionellen Behandlungsart um der Verständlichkeit willen erforderlich ist. Diese dürfte denn auch für fortgeschrittene Studenten mathematischer Fachrichtungen hinreichen, während von den Physikern wohl im allgemeinen erst graduierte Theoretiker aus dem Werk größeren Gewinn ziehen können. Den größten, wie uns scheint, aus den Darlegungen über Kontinuumsmechanik. (Grundprinzipien, sowie einiges über inkompressible einfache Flüssigkeiten, elastische isotrope Körper und — hervorhebenswert — über die sogenannten inhomogenen stoffgleichen (materially uniform) elastischen Körper.) Jenes Gebiet hat von den hier behandelten, was die fundamentalen Konzeptionen betrifft, denn ja auch die größten Fortschritte zu verzeichnen, abgesehen vielleicht von der allgemeinen Relativitätstheorie, deren Bearbeitung aber in vorliegendem Buche besonders stark im Konventionellen verbleibt, und wo nicht, eigenwillig in der Stoffauswahl ist. Wang kann eben seine Herkunft von der mathematischen Mechanik der Kontinua nicht verleugnen. Gleichwohl findet der Leser allerorten eine Fülle von Anregungen nicht zuletzt für den akademischen Unterricht. Was die sogenannte analytische Mechanik betrifft, so vertritt Verf. die Ansicht, daß der starre Körper ein vom Begriff des Massenpunkts unabhängiges, eigenständiges Konzept sei, und entwickelt dementsprechend die Lagrangesche Mechanik von holonomen, aus physikalischen Objekten beider Art bestehenden Systemen. Wir fanden jedoch nicht, ob und wie die Möglichkeit gegenseitiger Durchdringung der Körper ausgeschlossen werden soll. Sehr verständnisfördernd ist es, daß die Entwicklung der Lagrangeschen Theorie auf dem Begriff des Konfigurationsraumes aufbaut. Zum Anlaß weit umfangreicherer differentialgeometrischer Diskussionen nimmt Verf. die Hamiltonsche Version der Theorie. Die Analyse ihrer algebraischen Strukturen wäre allerdings von tieferer physikalischer Bedeutung gewesen. Als sehr sinnvoll erweist sich die Wahl des Ausganspunkts der ganzen Monographie: die vierdimensionale Raumzeitmannigfaltigkeit in ihrer Newtonschen Variante. Von hier aus kann sogleich eine saubere Definition der Bezugssysteme gegeben und die lex prima als Postulat der Existenz von Inertialsystemen verstanden werden. Von großem dialektischem Wert ist die (in der Mitte des zweiten Bandes befindliche) genauere Diskussion der Newtonschen Raumzeit im Hinblick auf die relativistischen Begriffsbildungen. Bei deren Einführung steht — der Sache sehr angemessen — wiederum die adäquate vierdimensionale Raumzeit, also zunächst die Minkowskische, am Anfang, nachdem das Erfordernis der speziellen Relativitätstheorie durch die Existenz eines Äthersystems motiviert worden ist, welche im Widerspruch zur Erfahrung von der Maxwellschen Theorie verlangt wird, wenn man sie auf dem Hintergrund der Newtonschen Raumzeitkonzeption behandelt. Verf. entwickelt die klassische Elektrodynamik unter der Voraussetzung eines solchen Bezugssystems. Enttäuschend sind die Ausführungen — sowohl die aus klassischer wie die aus relativistischer Sicht — über die Wechselwirkung des elektromagnetischen Feldes mit dem mechanischen Medium, wo doch der Autor als Experte für derartige Medien besonders zuständig wäre, die diesbezüglichen Fragen zu behandeln. In Anbetracht der ungewöhnlich umfangreichen Literatur über den elektromagnetischen Energie-Impulstensor und verwandte Gegenstände, halten wir seine Ausführungen für zu assertorisch und zu kurz. Die vierdimensionale Darstellung der Elektrodynamik macht in vorliegendem Werk ausgiebigen Gebrauch von dem Kalkül der äußeren Formen, als einem sachgemäßen, wiewohl in diesem Zusammenhang ungewohnten Mittel.

Dresden H.-G. Schöpf

Truckenbrodt, E., Fluidmechanik. Band 1: Grundlagen und elementare Strömungsvorgänge dichtebeständiger Fluide. 2., neubearb. und erweiterte Auflage. Berlin-Heidelberg-New York, Springer-Verlag 1980. XX, 371 S., DM 136,—. US $ 76.20. ISBN 3-540-09499-7

Mit dem Buch „Fluidmechanik" erscheint der erste Band einer jetzt zweiteiligen, völlig neu bearbeiteten und erweiterten 2. Auflage zu E. Truckenbrodts „Strömungsmechanik". Ziel und Aufgabenstellung der neuen Auflage sind unverändert geblieben: Das Werk will Lehrbuch für Studierende der naturwissenschaftlichen und technischen Fächer sein und dem berufstätigen Ingenieur und Physiker zum Nachschlagen dienen. Der dargestellte Stoff ist in der neuen Auflage straffer gegliedert und statt bisher in 8 nur in 6 Kapitel unterteilt. Kapitel 1 beschreibt die physikalischen Eigenschaften und Stoffwerte der Fluide wobei mechanische und thermische Einflüsse gleichrangig behandelt werden. Kapitel 2 beschreibt ausführlich die Grundgesetze der Fluid- und Thermo-Fluidmechanik bei ruhenden und strömenden Fluiden. Kapitel 3 und 4 befassen sich mit elementaren Strömungsvorgängen dichtebeständiger bzw. dichteveränderlicher Fluide, wobei die Fluidstatik als Sonderfall auftritt. Kapitel 5 faßt unter der Überschrift „Drehungsfreie und drehungsbehaftete Potentialströmungen" die Kapitel 6 und 7 der ersten Auflage zusammen, die Grenzschichtströmungen, bilden nun Kapitel 6. Wegen des erheblich angewachsenen Umfangs erfolgte eine Aufteilung auf zwei Bände, der vorliegende Band 1 enthält die Kapitel 1 bis 3.

Neu aufgenommen wurden insbesondere Grundgleichungen zu Thermo-Fluidmechanik und ihre Berücksichtigung in Kennzahlen und Stoffwerten sowie zahlreiche, sehr übersichtliche tabellarische Darstellungen. Die Einheiten wurden konsequent auf das internationale System umgestellt, die Formelzeichen weitgehend der Normung angepaßt. Sehr nützlich erscheint auch die erhebliche Erweiterung und Aktualisierung der jeweils an den Kapitelenden nachgewiesenen Grundlagen- und Spezialliteratur. Der Untertitel des Buches weist auf Grundlagen und elementare Strömungsvorgänge, diese Abgrenzung ist sinnvoll und muß akzeptiert werden, auch wenn dadurch eine Reihe von wichtigen Themen am Rande oder außerhalb der Betrachtung bleibt: Mehrphasenprobleme, Nicht-Newtonsche Medien, Methoden und Ergebnisse der Rechneranwendung u. a.

Bei der Darstellung des Stoffes wurden verstärkt die Grundprinzipien und Modellgleichungen in den Vordergrund gerückt und die technischen Anwendungsfälle den Grundprinzipien zugeordnet. Für jedes Stoffgebiet werden die grundlegenden strömungsmechanischen Vorgänge anschaulich erläutert, die mathematischen Beschreibungsmittel ausführlich zusammengestellt und in zahlreichen Abbildungen verdeutlicht sowie die Besonderheiten für unterschiedliche technische Anwendungen behandelt. Das Buch ist besonders geeignet, Grundlagenwissen eindringlich zu vermitteln und den nachschlagenden Praktiker vielfältig zu beraten und anzuregen.

Berlin H.-E. Peters

EINGEGANGENE BÜCHER

Bei der Schriftleitung sind folgende Bücher eingegangen (ausführliche Besprechung bleibt vorbehalten).

Meetz, K. / Engl, W. L., Elektromagnetische Felder. Mathematische und physikalische Grundlagen. Anwendungen in Physik und Technik. Hochschultext. Berlin-Heidelberg-New York, Springer-Verlag 1980. 192 Abb., XIII, 668 S., DM 69,–. US $ 40.80. ISBN 3-540-09597-7

Bullough, R. K. / Caudrey, P. J. (Eds.), Solitons. Topics in Current Physics 17. Berlin-Heidelberg-New York, Springer-Verlag 1980. XVIII, 389 S., 20 Abb., DM 76,–. US $ 44.90. ISBN 3-540-09962-X

Rath, H. J., Peristaltische Strömungen. Medizinische Informatik und Statistik 19. Berlin-Heidelberg-New York, Springer-Verlag 1980. VIII, 119 S., DM 25,–. US $ 14.80. ISBN 3-540-09999-9

Amazigo, J. C. / Rubenfeld, L. A., Advanced Calculus and Its Applications to the Engineering and Physical Sciences. New York-Chichester-Brisbane-Toronto, John Wiley and Sons 1980. VIII, 407 S., £ 10.25. ISBN 0-471-04934-4

Winfree, A. T., The Geometry of Biological Time. Biomathematics 8. Berlin-Heidelberg-New York, Springer-Verlag 1980. XIV, 530 S., 290 Abb., DM 59,50. US $ 32.80. ISBN 3-540-09373-7

Rautmann, R. (Ed.), Approximation Methods for Navier-Stokes Problems. Proceedings, Paderborn, Germany 1979. Lecture Notes in Mathematics 771. Berlin-Heidelberg-New York, Springer-Verlag 1980. XVI, 581 S., 202 Abb., DM 58,–. US $ 34.30. ISBN 3-540-09734-1

Bauer, K. W. / Ruscheweyh, S., Differential Operators for Partial Differential Equations and Function Theoretic Applications. Lecture Notes in Mathematics 791. Berlin-Heidelberg-New York, Springer-Verlag 1980. V, 258 S., DM 29,–. US $ 17.20. ISBN 3-540-09975-1

Izé, A. F. (Ed.), Functional Differential Equations and Bifurcation. Proceedings, São Carlos, Brazil 1979. Lecture Notes in Mathematics 799. Berlin-Heidelberg-New York, Springer-Verlag 1980. XXII, 409 S., 21 Abb., DM 43,50. US $ 25.70. ISBN 3-540-09986-7

Landau, L. D. / Lifschitz, E. M., Lehrbuch der Theoretischen Physik. Bd. 9: Statistische Physik, Teil 2 (von **Lifschitz, E. M. / Pitajewski, L. P.**). Berlin, Akademie-Verlag 1980. XI, 390 S., M 28,–. BN 7623184 (5436/IX). Übersetzung aus dem Russischen.

Sigloch, H., Technische Fluidmechanik, Hannover, Hermann Schrödel Verlag KG 1980, 328 S., 67 ausführliche Übungsbeispiele, DM 43,80. ISBN 3-507-94104-x

Schutz, B., Geometrical Methods of Mathematical Physics, Cambridge, Cambridge University Press 1980. £ 18.00 A H/C SBN 0-521-23271-6, £ 7.50 A P/B, SBN 0-521-29887-3

Gumowski, I. / Mira, C., Dynamique chaotique. Transformations ponctuelles, Tansition Ordre-Désordre. Toulouse, Cepadues Editions 1980. 480 S.,

Krause, F. / Rädler, K.-H., Mean-Field Magnetodynamics and Dynamo Theory. Berlin, Akademie-Verlag 1980. 271 S., 46 Abb., M 48,–. BN 761-626-7 (5948)

Jablonski, S. W. / Lupanow, O. B., Diskrete Mathematik und mathematische Fragen der Kybernetik. Mathematische Lehrbücher 31. Berlin, Akademie-Verlag 1980. Basel-Stuttgart, Birkhäuser Verlag 1980. 287 S., M 46,–. BN 7624021 (6401)

Landau, L. D. / Lifschitz, E. M., Theoretische Physik kurzgefaßt: II Quantentheorie. Berlin, Akademie-Verlag 1980. XII, 316 S., M 34,–. BN 7618481 (5879/II) (Übersetzung aus dem Russischen)

Elsner, N., Grundlagen der Technischen Thermodynamik. 4. bearbeitete und erweiterte Auflage. Berlin, Akademie-Verlag 1980. 641 S., 310 Abb., M 38,–. BN 7623758 (5913)

Olszak, W. / Suklje, L. (Eds.), Limit Analysis and Rheological Approach in Soil Mechanics. CISM Courses and Lectures 217. Wien-New York, Springer-Verlag 1978. VII, 292 S., 116 Abb., DM 52,–. S 358,–. ISBN 3-211-81511-2

Wesolowski, Z. (Ed.), Nonlinear Dynamics of Elastic Bodies. CISM Courses and Lectures 227. Wien-New York, Springer-Verlag 1978. V, 212 S., 38 Abb., DM 39,–. S 278,–. ISBN 3-211-81512-0

Kurzweil, J., Nichtabsolut konvergente Integrale. Teubner Texte zur Mathematik 26. Leipzig, BSB B.G. Teubner Verlagsgesellschaft 1980. 184 S., 2 Abb., M 18,–. BN 6660055

Eisenreich, G., Vorlesung über Funktionentheorie mehrerer Variabler. Teubner-Texte zur Mathematik 32. Leipzig, BSB B.G. Teubner Verlagsgesellschaft 1980. 136 S., M 14,–. BN 6660098

Sinha, S. C./Kirmser, P. G./Huang, C. L. D. (Eds.), Proceedings of the 16th Midwestern Mechanics Conference, Kansas, 1979. Developments in Mechanics 10. Kansas State University 1979. 324 S.

Besseling, J. F. / van der Heijden, A. M. A. (eds.), Trends in Solid Mechanics, Proceedings of the Symposium dedicated to the 65th Birthday of W. T. Koiter, Delft, 1979; Delft, Delft University Press and Sijthoff & Noordhoff International Publishers, 1979, Alphen aan den Rijn, XI, 246 S., Dfl. 90.00. $ 45.00 ISBN 90-286-0699-8

Tinhofer, G., Zufallsgraphen, Applied Computer Science 17, München-Wien, Carl Hanser Verlag 1980, 198 S., DM 34,–. ISBN 3-446-13172-8

NACHRICHTEN

Symposium „Numerische Mathematik und ihre Anwendungen"

Die Deutsche Akademie der Naturforscher LEOPOLDINA veranstaltet vom 10. bis 13. 3. 1982 in Halle (Saale) ein Symposium über „Numerische Mathematik und ihre Anwendungen", auf dem aktuelle Fragen der Numerischen Mathematik wie nichtlineare Probleme, inkorrekt gestellte Probleme, Optimierungsprobleme und Fehlerabschätzungen für Lösungen verschiedener Gleichungstypen sowie für Eigenwertprobleme im Mittelpunkt stehen werden und außerdem der Einsatz numerischer Methoden und der modernen Rechenautomaten zur Bearbeitung verschiedener physikalisch-technischer Anwendungen.

Anfragen sind an Prof. Dr. H. Schwetlick, Sektion Mathematik der Martin-Luther-Universität, DDR-4020 Halle (Saale) zu richten.

Fachtagung Maschinendynamik 1982 in Karl-Marx-Stadt

Vom 8. bis 10. Juni 1982 veranstaltet die Technische Hochschule Karl-Marx-Stadt gemeinsam mit dem DDR-Komitee der IFToMM, Fachunterausschuß Maschinendynamik der KDT und dem KDT-Bezirksverband Karl-Marx-Stadt die Fachtagung „Maschinendynamik".

Ziel dieser Tagung mit internationaler Beteiligung ist die Vermittlung des Kenntnisstandes der Dynamik von Werkzeug- und Verarbeitungsmaschinen und die Anregung des Erfahrungsaustausches.

Folgende Problemkreise werden behandelt:

- Modellbildung und Kennwertermittlung
- Allgemeine Berechnungsverfahren
- Dynamik von Werkzeugmaschinen
- Dynamik von Verarbeitungsmaschinen.

Interessenten wenden sich an folgende Adresse:

Technische Hochschule Karl-Marx-Stadt Sektion Maschinen-Bauelemente, LS Dynamik (11124) DDR 9001 Karl-Marx-Stadt, PSF 964

Beauftragter für Organisation: Dr.-Ing. Waldeck, Tel. 668340 Tagungsbeiträge (Vortragsdauer 15 Minuten) sind bis zum 15. 8. 1981 anzumelden.

FSC
www.fsc.org
MIX
Papier | Fördert gute Waldnutzung
FSC® C083411

Zeitfracht Medien GmbH
Ferdinand-Jühlke-Straße 7
99095 Erfurt, Deutschland
produktsicherheit@kolibri360.de